WABASH CARNEGIE P LIB
1230 9100 198 719 5

SCIENCE

FOUNDATIONS

Quantum Theory

SCIENCE FOUNDATIONS

The Big Bang
Cell Theory
Electricity and Magnetism
Evolution
The Expanding Universe
The Genetic Code
Germ Theory
Gravity
Heredity
Kingdoms of Life
Light and Sound
Matter and Energy
Natural Selection
Planetary Motion
Plate Tectonics
Quantum Theory
Radioactivity
Vaccines

SCIENCE FOUNDATIONS

Quantum Theory

PHILLIP MANNING

Science Foundations: Quantum Theory

Chelsea House
An imprint of Infobase Learning
132 West 31st Street
New York, NY 10001

Library of Congress Cataloging-in-Publication Data
Manning, Phillip, 1936–
Quantum theory / by Phillip Manning.
p. cm. — (Science foundations)
Includes bibliographical references and index.
ISBN 978-1-60413-295-3 (hardcover)
1. Quantum theory. I. Title.
QC174.12.M3534 2011
530.12—dc22 2010029479

Chelsea House books are available at special discounts when purchased in bulk quantities for businesses, associations, institutions, or sales promotions. Please call our Special Sales Department in New York at (212) 967-8800 or (800) 322-8755.

You can find Chelsea House on the World Wide Web at http://www.infobaselearning.com

Text design by Kerry Casey
Cover design by Alicia Post
Composition by EJB Publishing Services
Cover printed by Yurchak Printing, Landisville, Pa.
Book printed and bound by Yurchak Printing, Landisville, Pa.
Date printed: June 2011
Printed in the United States of America

10 9 8 7 6 5 4 3 2 1

This book is printed on acid-free paper.

All links and Web addresses were checked and verified to be correct at the time of publication. Because of the dynamic nature of the Web, some addresses and links may have changed since publication and may no longer be valid.

Contents

The End of Physics

"Physics is a branch of knowledge that is just about complete. The important discoveries, all of them, have been made. It is hardly worth entering physics anymore." These sentiments were expressed in 1875 by a professor who was the head of the physics department at the highly respected University of Munich. The person who asked for the professor's opinion was a 17-year-old boy named Max Planck.

Twenty-five years later, Planck would prove the professor wrong. In fact, he would show that the professor was about as wrong as one can be. Because Max Planck's work not only led to new discoveries, it revolutionized physics. Planck's most daring idea was to hypothesize that energy was not continuous but came in tiny packets called "**quanta**." This simple idea opened the door to the **quantum theory**.

It is easy today to laugh at the physics professor who proclaimed that all the important discoveries had been made. At the time, though, it was not a bad guess. The reason for this attitude was the outstanding successes of **classical physics** during the previous 200 years. Although many people contributed to the flowering of classical physics, its development was dominated by three brilliant men: Galileo Galilei, Isaac Newton, and James Clerk Maxwell.

GALILEO GALILEI (1564–1642)

Many historians of science regard Italian physicist Galileo Galilei's work in astronomy as his greatest contribution to science. After all, he was the first person to spot the moons of Jupiter. He also observed the phases of Venus, which proved (at least to some people) that Venus was orbiting the Sun. And, using the newly developed telescope, he pointed out that the moon was not a perfect sphere, composed of some uniform, celestial material as the great Greek philosopher Aristotle had said. Galileo disposed of that ancient notion with one sentence: The moon is, he wrote, "not robed in a smooth and polished surface [but is] . . . rough and uneven, covered everywhere, just like Earth's surface, with huge prominences, deep valleys, and chasms."

Eventually, Galileo's insistence that Earth revolved around the Sun, rather than vice-versa, as the Catholic Church preached at the time, got him in trouble. In an oft-told story, Galileo got down on his knees before the Inquisition of Rome and recanted that belief.

So important was Galileo's work in astronomy and the dramatic tale associated with it, that it is easy to overlook his contributions to classical physics. Galileo was a pioneer in the field of mechanics, the study of how objects behave when forces act on them.

The laws of **motion** Galileo learned in school originated with Aristotle. Fundamental to Aristotle's laws was the idea that objects seek their natural place. Rocks fall toward the ground because the ground is where rocks come from. How fast do things fall? Aristotle said that heavy objects fall faster than lighter ones in proportion to their weights. For some situations, this is a nice common sense conclusion: Feathers do fall slower than rocks. Still, why some Greek or early European did not cut a brick in half and drop one of the halves alongside a whole brick to see if they hit the ground at the same time is hard to fathom.

Finally, Aristotle held that an object moved at a speed proportional to the force applied to it. Again, under certain conditions, this law makes sense. Push a heavy box across a concrete floor, and the harder you push, the faster the box moves. Stop pushing and the box stops moving. Aristotle's ideas stood for almost 2,000 years, until Galileo dismantled them.

Galileo was a practical man who tested his ideas with experiments. To test Aristotle's thesis about the speed of falling bodies, he dropped a cannonball and a much lighter musket ball from a high place (although probably not from the Leaning Tower of Pisa, as legend has it). The two objects fell at equal speeds. Over time,

Galileo became convinced that if air resistance were eliminated, all bodies, from feathers to cannonballs, would fall at the same speed.

Galileo also doubted Aristotle's second law of motion, namely that a **force** was required to keep a body in motion. Instead of pushing a box across the floor, try pushing it across an ice rink and you will see that the box does not stop moving when you stop pushing. These ideas led Galileo to perform a series of experiments that involved rolling balls down inclined planes.

Galileo noted than when two inclined planes faced one another, the ball would roll down one and up the other. Using smoother planes and rounder balls, Galileo saw that a ball released at certain height on one inclined plane would roll to almost that same height on the facing inclined plane. This led Galileo to conclude that if friction was entirely eliminated, the ball would roll to exactly the same height on the opposite inclined plane.

This observation produced a key insight: What would happen in a frictionless system if a ball rolled down an inclined plane onto a perfectly flat plane? It would, he reasoned, continue to roll forever at the same speed as long no other force acted on it. This property of matter is called **inertia**, and it is the property that causes objects to resist any change in motion. A body at rest will stay at rest unless acted upon by a force. A body moving in a straight line at constant speed will continue to do so unless a force is applied.

Although Galileo's contributions to astronomy were crucial to that branch of science, this fundamental insight into how and why objects move might have been even more important. For one thing, it helped Isaac Newton develop his own laws of motion, laws that would revolutionize physics and the way people regarded the world we inhabit.

ISAAC NEWTON (1642–1727)

Isaac Newton was born on Christmas Day, 1642, the year Galileo died. His birthplace was Woolsthorpe Manor, a farmhouse in a tiny village in the east of England. Two of the things one can say for sure about Newton are that he was a genius and a very difficult man. He invented the reflecting telescope, and his basic design is still used in

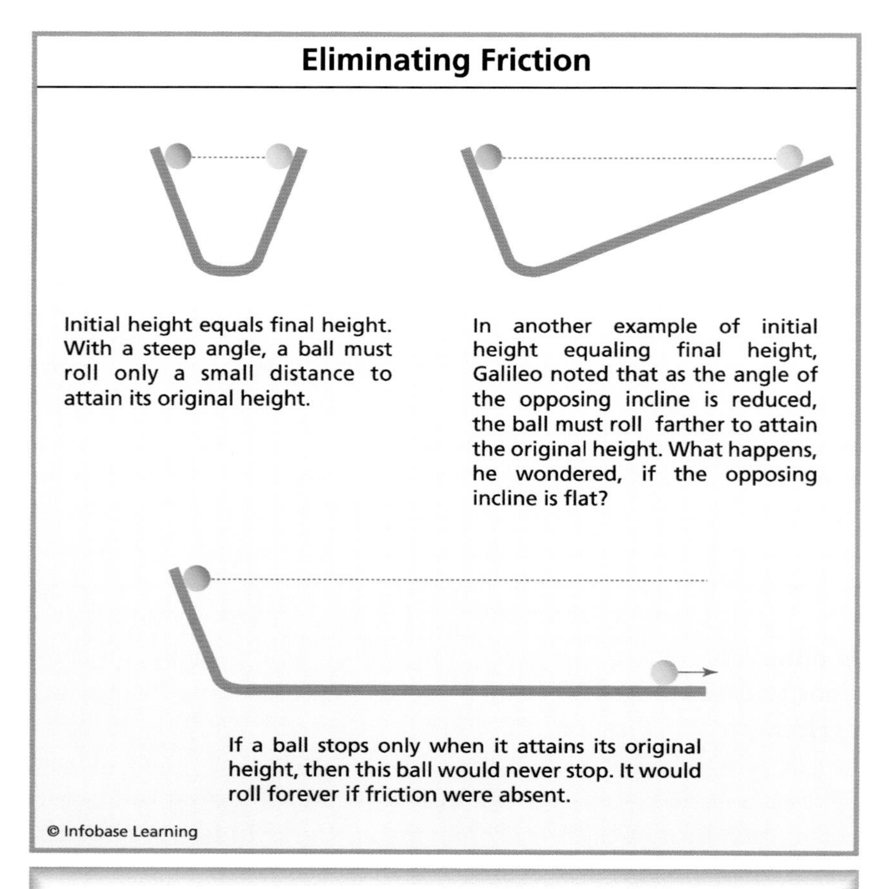

Figure 1.1 Galileo's experiments with friction yielded a crucial observation.

almost every observatory in the world; he advanced the science of optics enormously and discovered that white light is a blend of all the colors in the visible spectrum; he invented calculus, a mathematical methodology used routinely by engineers and scientists today; and he developed the laws of mechanics and gravitation that stood unchallenged for over 200 years and are still handy today for everything from dam building to space flight.

Many of Newton's ideas were conceived in 1665, the year that the Great Plague cut a deadly swath through England. It was to be the last—and the worst—epidemic of the bubonic plagues that had periodically depopulated England beginning with the Black Death of 1348. The death rates in steamy London during the hot summer of 1665 reached as high as 8,000 victims a week. Anyone who could leave London (or any other infected town) did. The University of Cambridge was effectively shut down. Newton, a Fellow of the College, retired to his farm in Woolsthorpe for 18 of the most productive months in the history of science.

Back at Woolsthorpe, Newton worked alone and kept notes, but submitted nothing for publication. This continued the practice he had started as a student at Cambridge and that he would follow, with few exceptions, for the rest of his life. During this time, Newton did much of his groundbreaking work on calculus, the reflecting telescope, and optics. He also began to formulate his law of gravity. In those months at Woolsthorpe, Newton said later, "I was in the prime of my age for invention and minded mathematics and philosophy more than at any time since."

In fact, Newton was a driven man, a workaholic. He was also one of the great geniuses of all time. In addition to his pioneering work in optics and mathematics, he made even more important contributions to classical mechanics, much of which he worked out during his miracle months at Woolsthorpe, but did not publish.

Edmund Halley (for whom the famous comet is named) knew Newton as well as anyone. One day, he challenged his friend with a question: Why do the planets move as they do? Newton said he knew the answer and promised Halley that he would write up his ideas. After much hard work and many delays, he finished the *Principia*, which was published in July 1687. It is arguably the most important science book ever written.

Newton begins by defining terms: **mass**, **motion**, and **force**, among others. He follows this with three of the most basic laws of physics, now known as Newton's laws of motion.

1. Every body perseveres in its state of being at rest or of moving uniformly straight forward, except insofar as it is compelled to change its state by forces impressed.
2. A change in motion is proportional to the motive force impressed and takes place along the straight line in which that force is impressed. This law is usually stated as an equation.

$$F = ma$$

Here, F is the applied force, m is the mass of the object, and a is the acceleration that the force imparts to it.

3. To any action there is always an opposite and equal reaction; in other words, the action of two bodies upon each other are always equal and always opposite in direction.

This is genius. People had been puzzling over how and why things moved for millennia. Galileo's concept of inertia had foreshadowed Newton's first law, but no one else had even come close to getting at the entire truth about the interaction between forces and matter before Newton. Here were the answers to nature's most fundamental questions summarized in three sentences. As science writer Timothy Ferris put it: "Never before in the history of empirical thought had so wide a range of natural phenomena been accounted for so precisely, and with such economy."

Newton's impact went far beyond the scientific world. His work led directly to the French Enlightenment and changed forever the way much of humankind views the world. The physical world does not follow the rules of capricious gods or kings; it follows the laws formulated by Newton.

Newton died in 1727 at the age of 84. He was buried with great honor in Westminster Abbey, where he lies alongside England's most famous politicians and poets, kings, and philosophers. Alexander Pope wrote an epitaph that summarized Newton's contributions: "Nature and Nature's laws lay hid in night/God said, *Let Newton be!* And All was *Light*."

Many other prominent people heaped praise on Newton, but Newton himself had earlier judged his contributions more temperately. "If I have seen further," he wrote, "it is by standing on the shoulders of giants." Another quote is equally, if uncharacteristically, modest: "I don't know what I may seem to the world, but as to myself, I seem to have been only like a boy playing on the sea-shore, and diverting myself in now and then finding a smoother pebble or a prettier shell than ordinary, whilst the great ocean of truth lay all undiscovered before me."

These two quotations nicely characterize the scientific enterprise. Only in the rarest cases can one separate out the contributions of one individual and say he or she alone accomplished some

Newton's Apple

Although Newton's laws of motion are models of simplicity, using those laws to solve real-life problems can be tricky. Consider, for instance, the simple problem of young Newton sitting under an apple tree at Woolsthorpe, maybe sleeping, maybe formulating his law of gravitational attraction. Meanwhile, above him, an apple hangs from a tree.

Gravity is pulling the apple downward, and the tree branch is exerting an equal force on the apple in the opposite direction, just as Newton's third law of motion dictates. First, how big is the force acting on the apple? This is easily calculated from Newton's second law:

$$F = ma$$

Yet think about it: The apple is not moving. Therefore, its acceleration equals 0 and

$$F = (m)(0)$$

and therefore

$$F = 0$$

Still, the force of gravity must be pulling on the apple, even though the second law insists that the forces acting on it are 0. How can these two statements be reconciled? The answer lies in a sometimes overlooked aspect of Newton's laws. The term F, in the equation $F = ma$, really means net force. And because the force of gravity is exactly equal to the restraining force of the branch holding the apple, the net force is 0.

$$F_{net} = F_{gravity} - F_{restraint} = 0$$

And because the net force acting on the apple is 0, it stays in the tree rather than falling and hitting Newton on the head.

particular thing. All science depends on what went before. And all scientists resemble Newton playing on the seashore—and, every now and then, they find a smoother pebble.

JAMES CLERK MAXWELL (1831–1879)

James Clerk Maxwell is the last of the three giants of classical physics whose insights made Max Planck's professor suspect that physics was near an end. Maxwell is less well known than Galileo or Newton, but to scientists, he is a towering figure.

Maxwell is celebrated today by (among other things) a geeky T-shirt with four daunting equations on the front. These are **partial differential equations**, and they will not be discussed here because understanding them requires advanced mathematics. However, they are beautiful to those who can follow the math.

Those equations establish the relationship between electricity and magnetism. Using them, Maxwell predicted the existence of **electromagnetic waves** and calculated that such waves travel at the speed of light. In fact, he concluded, light itself is an electromagnetic wave.

It's hard to pin down the physical nature of electromagnetic waves. One can start by visualizing a stationary electric charge that produces an electric field. Then, think of a moving charge that also produces a magnetic field. However, the moving magnetic field gives rise to an electric field. Thus, a moving electric charge produces a wave that contains both electric and magnetic components (Figure 1.3).

The man who united electricity and magnetism in four concise equations was born in Edinburgh, Scotland, into a prominent and wealthy family. He followed the footsteps of Isaac Newton and many other outstanding scientists to college at Cambridge. Maxwell later returned to Cambridge as a professor to set up the Cavendish physics laboratory, which soon became one the great centers for physics research in the world.

Maxwell made many contributions to science, including taking the world's first color photograph. He also had a hand in the development of statistical mechanics. However, more than his many other contributions to science, his discovery of the nature of light

Figure 1.2 A young James Clerk Maxwell, pictured during his time at the University of Edinburgh in Scotland, holds the color wheel he invented. Besides his work with electricity and magnetism, Maxwell is known for creating the wheel to demonstrate how all colors could be made by a mixture of three primaries: vermillion (scarlet), emerald green, and ultramarine blue. When the wheel is spun, the three colors blend together to create the new color.

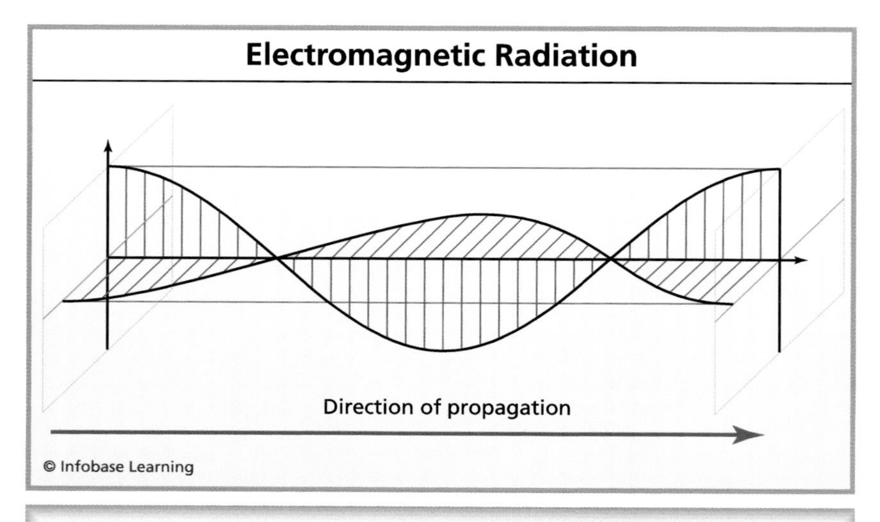

Figure 1.3 In this image, a moving electric charge creates varying electric and magnetic fields. Blue represents the magnitude of the electric field, while red represents the magnitude of the magnetic field.

was the one that landed him among the titans of classical physics. This discovery led to the understanding that light waves were merely one small region of wavelengths in a much wider spectrum of electromagnetic radiation. This, in turn, led to the development of wireless telegraphy, and, later, to radio, radar, and television.

Maxwell died in Cambridge at age 48 of abdominal cancer. His enormous legacy may be unknown to all but a few, but to those few, Maxwell was one of the greatest scientists. When Albert Einstein visited Cambridge, long after Maxwell's death, one of his colleagues remarked that "You have done great things but you stand on Newton's shoulders." "No," replied Einstein, "I stand on Maxwell's shoulders."

THE PREDICTABLE UNIVERSE

After Newton—building on Galileo's ideas about inertia—proposed his three laws of motion, plus the law of gravitational attraction, astronomers could predict accurately the motions of the planets for the first time in history. This led to the idea of a clockwork universe:

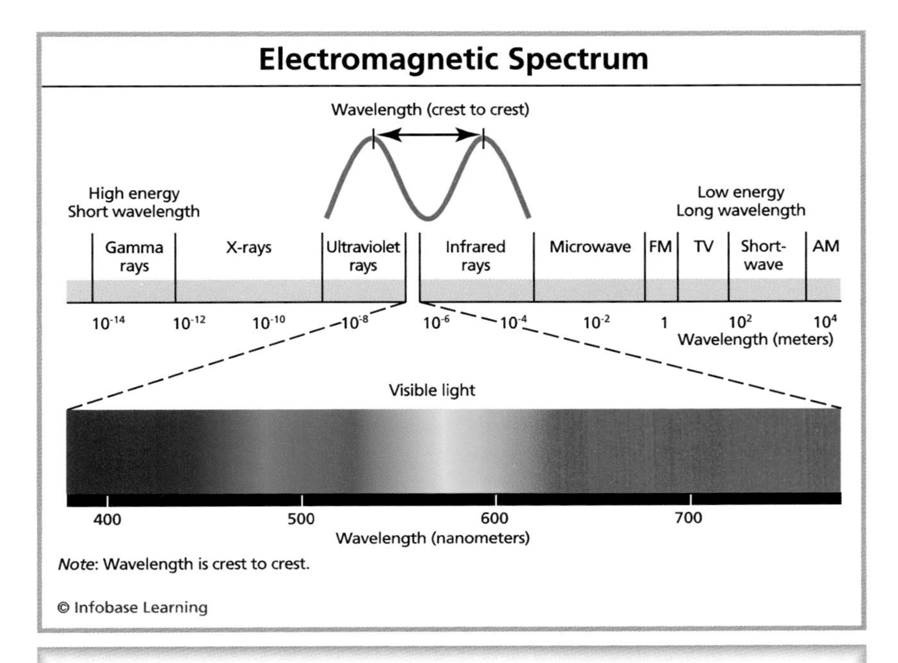

Figure 1.4 Higher energy corresponds to a shorter wavelength and higher frequency, while lower energy corresponds to longer wavelengths and lower frequency.

a universe that God set in motion but that subsequently followed the laws of Newton. One example of this concept was the invention of a mechanical device known as an orrery, which showed the relative positions of the known planets of the solar system. With this device, one could move, say, Earth and watch the rest of the planets and the Moon rearrange themselves in accordance with Newton's laws.

The idea of a clockwork universe combined with Maxwell's recent discovery of the relationship between electricity, magnetism, and light may have been behind Max Planck's professor's conclusion that all the big advances in physics were complete.

Fortunately, Max Planck did not believe him. He took up physics and, by 1900, he was a respected professor at the prestigious University of Berlin, working on what was known as the **blackbody** problem.

A blackbody is an idealized object that emits and absorbs radiation. When heated, the intensity and total energy of its emitted

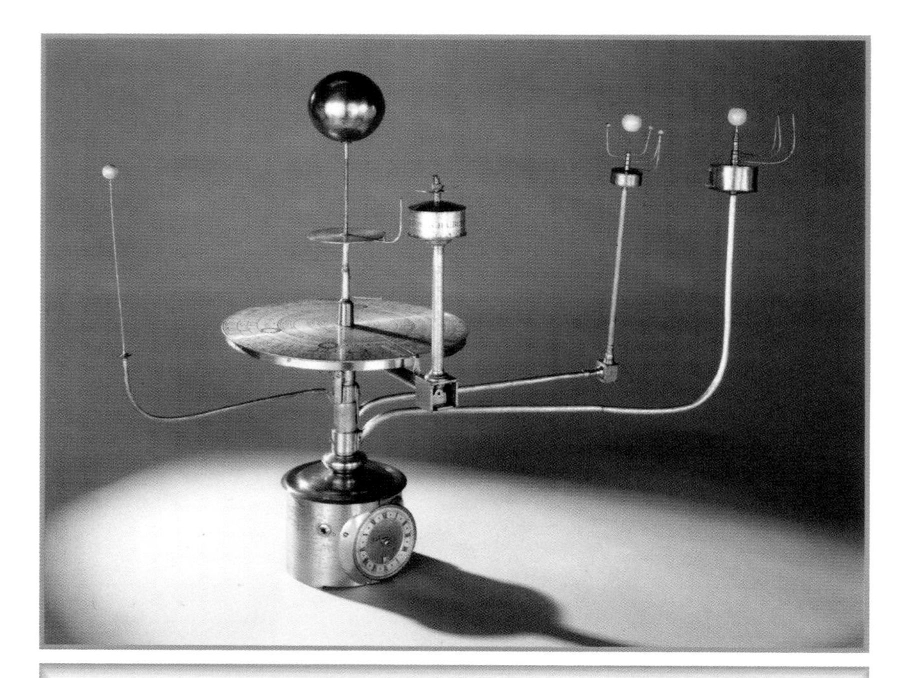

Figure 1.5 British clockmaker Charles Butcher made this orrery in 1733 for Lord Trevor, Baron of Bromham. An eight-day clock rotates the planets. The calendar ring is 4.5 feet (1.4 m) in diameter and has painted metal figures for the signs of the zodiac.

radiation increases. Also, the wavelength of the emitted radiation shifts from longer to shorter wavelengths: from infrared to visible to ultraviolet and beyond. Most solid objects—a fireplace poker, for example—closely mimic blackbodies. When a poker is heated in a fire, at first its color does not change, although heat is radiating from it in the form of infrared radiation. This radiation is invisible to our eyes, but we can feel it with our hands. Heat the poker to a higher temperature and it will turn red. The radiation becomes visible because the poker is emitting higher energy waves, which our eyes can detect. This is true for any blackbody. Higher temperatures produce more energetic radiation of higher intensity (Figure 1.6).

This result was experimentally determined in the nineteenth century, but it could not be explained by the physics of Galileo, Newton, or Maxwell. Try as they might, physicists could not come up with an equation that gave the results observed in blackbody radiation.

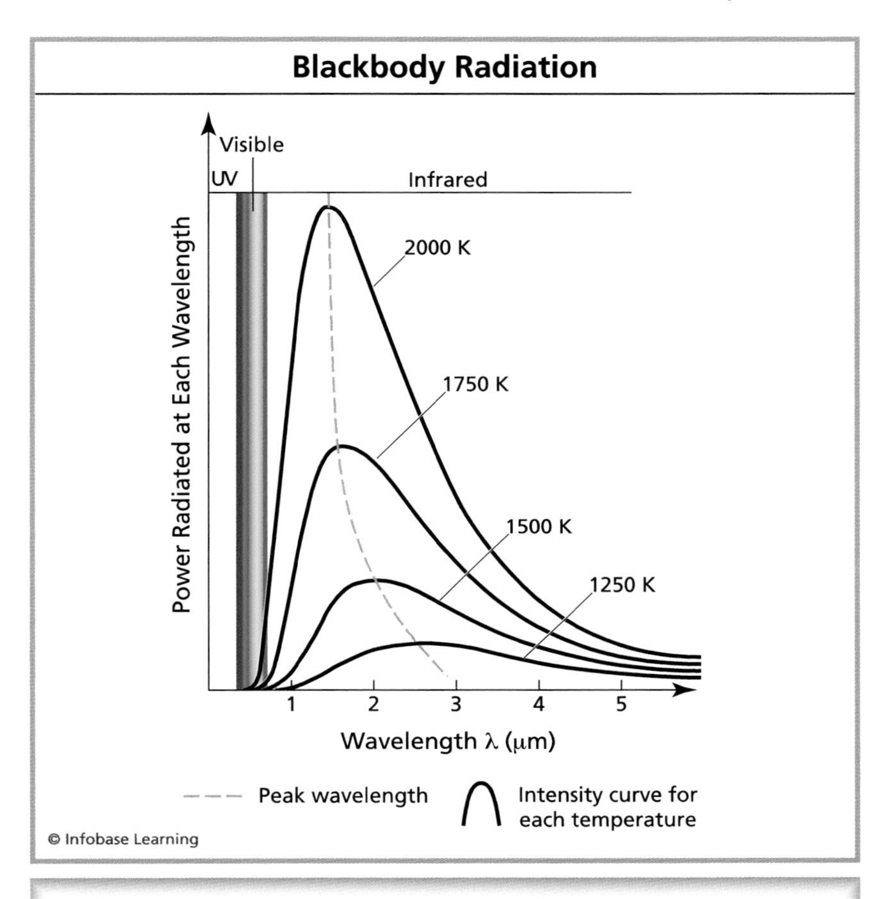

Figure 1.6 As its temperature increases, a blackbody produces radiation that is more energetic (indicated by the x-axis of this graph) and more intense (indicated by the y-axis).

Max Planck was a highly disciplined man with a highly disciplined mind that was steeped in classical physics. A great disciple of his field, he would be the one to solve the blackbody problem. His solution would lead the way to a new physics—a physics governed by the quantum theory—that would sweep away the certainties of the clockwork universe. The development of that theory would also put to rest the fears of Planck's professor that "the important discoveries, all of them, have been made." In fact, many shiny pebbles lay on the seashore awaiting discovery by the scientists of the twentieth century.

2

The Revolution Begins

Max Planck was an unlikely revolutionary. The descendent of generations of conservative German academics who specialized in theology or law, Planck fit the same mold in all but his choice of profession. Virtually every photograph shows him dressed in a formal dark suit with properly knotted tie and vest. According to his biographer J.L. Heilbron, "Respect for law, trust in established institutions, observance of duty, and absolute honesty . . . were the hallmarks of Planck's character." Because he is a pivotal figure in the history of science, let's learn a bit more about this respectful man with a buttoned-down mind who upended the long-standing laws of physics.

MEET MAX PLANCK

Max Planck was born in Germany in 1858 and lived to be almost 90 years old. He was a respected professor of physics at the University of Berlin and, later, president of the Kaiser Wilhelm Institute, the most prestigious scientific position in Germany. During his long career, he was awarded many medals and prizes, including, in 1918, the most coveted one of all—the Nobel Prize in Physics. What should have been a happy, fulfilled life, however, was tempered by war. During Planck's lifetime, Germany entered and lost two world wars.

Figure 2.1 German physicist Max Planck pioneered quantum mechanics when, in a 1901 paper describing blackbody radiation, he proposed that radiation must be emitted or absorbed in energy packets called quanta.

The years leading up to the World War II were an especially difficult time for Planck and German science. Planck and Albert Einstein, his Jewish friend and colleague at the University of Berlin, witnessed the rise of Hitler and the Nazi Party with dismay. By the 1920s, Einstein, who had opposed Germany's entrance into World War I, was being regularly heckled by the anti-Semitic Nazis. He had to pull out of some scientific conferences because of threats to his life. Planck was appalled at what was happening. Germany, he said, should not allow "a band of murderers, who go about their business in the dark, dictate the scientific program of a purely scientific body."

By the end of 1932, the political climate for Jews in Germany was so bleak that Einstein left for the United States, never to return. Planck was sympathetic toward Einstein and other Jewish scientists who were being forced from their positions in government and universities. However, he elected to stay in Germany to try to quietly fight the rising tide of anti-Semitism and Nazism that was sweeping through the land.

Planck's old friend Einstein was unhappy with his refusal to speak out directly against the Nazis. Einstein, however, had no strong attachment to Germany, while Planck came from a long line of solid German citizens and was deeply involved in the country's culture and institutions. He decided, according Heilbron, "to stay and salvage rather than to run and gesture."

This may not have been the best decision. When the Allies bombed the suburb where Planck lived, he lost everything—his house, his library, his papers. Then, his son Erwin—his last living offspring of the four sons and daughters he had fathered—was accused of participating in the 1944 plot to assassinate Hitler. Planck pulled every string to save him but, nevertheless, Erwin was executed in January 1945. "My sorrow," Planck wrote, "cannot be expressed in words."

Despite the tragedies in his personal life, Planck was an optimistic man, and it showed in his approach to science. His objective was to understand the foundations of physics, to probe nature at the deepest level. His positive outlook led him to believe that he could make important contributions to science by pursuing that difficult goal. This ambition led him to search for a better theory of **oscillators**, which, in turn, led him to the blackbody problem.

ENTER THE QUANTUM

One class of theories that tried to explain the spectrum of blackbody radiation was based on the notion that blackbodies are composed of tiny oscillators. These oscillators produce a continuum of electromagnetic waves—analogous to the sound waves that you get when you pluck a violin's strings. Physicists developed several models involving oscillators to predict the spectrum of blackbodies. Yet, either the spectrum they predicted did not fit the experimental data or the model had a questionable theoretical foundation, both of which Planck found unsatisfying. Further, at the high energies of the ultraviolet range, prediction and experiment were wildly different in some models. So serious was this breakdown of the laws of physics that scientists later labeled it "the ultraviolet catastrophe."

Planck was working on another problem involving oscillators and blackbodies when he stumbled onto the idea that led to **quantum mechanics**. He had been trying to count the number of ways energy can be distributed in a set of oscillators. To find this number, he had to count the oscillators vibrating at a given wavelength. However, oscillators must be counted with integers. There is no such thing as a fractional oscillator. Thus, as Planck himself later wrote, the total energy of any system of oscillators must "be expressible as a sum of identical energy elements...." Planck later came up with the term used today to designate these tiny bits of energy. He called them *quanta*.

Planck presented his revolutionary paper on blackbodies to the German Physical Society in December 1900. His big idea swept away a basic tenet of classical physics, namely that energy is continuous. Until Planck came along, energy was thought to act like most other things—pizza, for instance. One can eat as much or as little pizza as one wants. Yet, if Planck's idea of quanta were applied to pizza eating, then that enjoyable pastime would be no longer be continuous. Quantum rules would insist that pizza comes in slices and that you must eat all or none of a slice. When the quantum concept was applied to the subatomic world, it caused a paradigm-busting change in the way scientists view the world—although the new rules had no effect on how pizza is actually eaten.

The equation that mathematically summarized this quantum insight is known today as Planck's equation. It is written as

$$E = nhf$$

where E is the energy of an oscillator in the blackbody (the ever-cautious Planck did not call the oscillators atoms, but it was clear that he was referring to atoms), *f* is its frequency of oscillation, *n* is an integer (0, 1, 2, 3 . . .) and *h* is a very small number. This number is known today as Planck's constant. It is usually represented using **scientific notation** as 6.6×10^{-34} joule-seconds. (A joule is the International System of Units or **SI unit** of work. Abbreviated J, it is equal to 0.24 calories.) In decimal form, Planck's constant looks like this:

$$0.00000000000000000000000000000000066 \text{ J}\bullet\text{s}$$

Thus, for a single atom, the smallest quantity of energy that it could absorb or emit would be *hf*. An atom could emit or absorb radiation of frequency *f* or 2*f* or 3*f*, but not 0.5*f* or 1.8*f*. When Planck used his equation to calculate the spectrum of blackbody radiation, he came up with a result that agreed perfectly with the experiment. For the first time, the spectrum shown in Figure 1.5 could be theoretically derived, thus eliminating the ultraviolet catastrophe. More importantly, Planck had unlocked the door to quantum mechanics, because this simple equation forms the basis of quantum theory.

Today, historians of science argue over the value of Planck's contribution to quantum theory. He was a deeply conservative man and a staunch believer in classical physics. After he published his seminal paper on blackbodies, Planck confessed that the quantization of energy was "a purely formal assumption and I really did not give it much thought . . ." It was, he suspected, a nice mathematical trick that solved a difficult problem. But was it reality? Was energy really discontinuous?

Planck's uncertainty about the existence of his newly proposed bits of energy would be resolved a few years after his original presentation. In 1905, an up-and-coming physicist who was working as a file clerk in Switzerland published a paper on the **photoelectric effect** that explained clearly the discontinuous nature of light. That clerk was Albert Einstein, and he presented a beautifully clear analysis of the photoelectric effect that erased any doubt physicists might have had about the existence of Planck's quanta. Planck himself became a believer, too, prophesying in a 1911 lecture that "the hypothesis of quanta will never vanish from the world." Never is a long time, but he has been right so far.

Using Planck's Equation

With Planck's equation, $E = nhf$, one can calculate the energy in a single quantum of light: Simply choose a color, find the associated wavelength, and calculate away. Take blue, for example. Figure 1.3 shows that blue light has a wavelength of about 475 nanometers (nm). The equation that relates the frequency of a wave to its wavelength is

$$f = v/\lambda$$

where f is the frequency of the wave (i.e., the number of complete waves that pass a given point in 1 second); v is the velocity of the wave in meters per second, and λ is the wavelength, the distance between the peaks (or troughs) of the wave as measured in meters. To calculate the frequency of blue light, divide the velocity of the wave, the speed of light, by the wavelength. (The velocity of light in a vacuum is usually designated by the letter c.)

$$v = c = 300{,}000{,}000 \text{ m/s}$$
$$c = 3.0 \times 10^8 \text{ m/s}$$
$$\lambda = 475 \text{ nm}$$

or

$$\lambda = 4.75 \times 10^{-7}\ m$$

Therefore,

$$f = \frac{3.0 \times 10^8 \text{m/s}}{4.75 \times 10^{-7} \text{m}}$$

$$f = 6.3 \times 10^{14} \text{ cycles/s}$$

Plugging f into the Planck equation:

$$E = nhf$$

where

$$n = 1,\ h = 6.6 \times 10^{-34} \text{ J•s}$$

gives

$$E = (6.6 \times 10^{-34} \text{ J•s})\ (6.3 \times 10^{14} \text{ cycles/s})$$
$$E = 4.16 \times 10^{-19} \text{ J}$$

(continues)

(continued)

Clearly, a single quantum of blue light has only a tiny bit of energy. However, one can get a better feel for this small quantity by changing the units in which it is expressed. A more appropriate unit is the **electron** volt (eV). One electron volt is the **kinetic energy** received by one electron passing through an electric potential of 1 volt. The relationship between joules and electron volts is

$$1\text{eV} = 1.6 \times 10^{-19}\text{ J}$$

Thus, one can convert the energy of a quantum of blue light from joules to electronic volts as follows

$$E = \frac{4.16 \times 10^{-19}\text{J}}{1.6 \times 10^{-19}\text{J/eV}}$$

$$E = 2.6\text{ eV}$$

The energy of electromagnetic radiation varies widely. Table 2.1 gives the energy of a single quantum of radiation from several regions of the electromagnetic spectrum.

Table 2.1

Type of Radiation	Approximate Wavelength (m)	Energy of One Quantum (eV)
X-rays	10^{-10}	12,500
Blue-green light	5×10^{-7}	2.5
Infrared	10^{-5}	0.125
Radar	10^{-2}	0.000125
AM radio	10^{4}	0.000000000125

3

Einstein Sees the Light

Science advances by many different paths: a brilliant experiment, a flash of insight, or years of tedious lab work. Another route to a scientific breakthrough starts with an unanticipated outcome, a surprise experimental result. Some scientists ignore these anomalies as they focus single-mindedly on their goal. Others dismiss oddball results, figuring that they arose from experimental error. Other scientists, however, will not turn away when an experiment gives unexpected results. They become determined to track down the source of the discrepancy. Physics professor Heinrich Hertz was one of the latter, and is a spectacular example of how pursuing the odd bit of data sometimes pays off big.

HERTZ DISCOVERS THE PHOTOELECTRIC EFFECT

In 1861, James Clerk Maxwell calculated that the speed of electromagnetic waves was the same as the experimentally measured speed of light. Therefore, he concluded, light must be a form of electromagnetic radiation. This was one of the great discoveries in the history of science.

Light could be detected by the eye, but if it was an electromagnetic wave, then an electric apparatus should also be able to generate and detect the waves. No matter how brilliant Maxwell's ideas were, hardheaded scientists demanded experimental verification. That verification would be provided by Heinrich Hertz.

Hertz was born in Hamburg, Germany, in 1857. By age 30, he was an aspiring physics professor. The main thrust of his research was to generate and detect electromagnetic radiation. He was a firm believer in Maxwell's theory and wanted to confirm it experimentally.

His idea was to generate electromagnetic waves with a primitive spark-gap device similar to the transmitters that were later used in the earliest radios. The even-more-primitive receiver was a circle of copper wire with an air gap in it. The wire on one side of the gap tapered to a point. The wire on the other side was tipped by a small, metal sphere. The size of the gap between point and sphere was very small and could be adjusted with a screw mechanism. The receiver was located a few meters away from the transmitter and was not connected to it. If electromagnetic radiation was generated by the transmitter, Hertz reasoned, then it should induce a current in the receiver, creating a tiny but visible spark across the receiver's air gap.

Amazingly, the difficult experiment worked. Hertz generated and detected radio waves, just as Maxwell's famous theory predicted. However, working with the apparatus was a tricky process. The spark emitted by the receiver was tiny and hard to see. To solve that problem, Hertz occasionally enclosed the spark in a dark box so it would be easier to see. When it was enclosed, Hertz noted, "the maximum spark-length became decidedly smaller" than when it was exposed to light.

Many of us might have responded by saying, "So what?" Hertz did not. To him, "A phenomenon so remarkable called for closer investigation." He next exposed the spark to ultraviolet light. Hertz discovered that the spark would jump a wider gap than it would when the receiver was in the dark or exposed to visible light. Somehow, shining light on the metal enhanced the current flowing in the receiver, creating a more vigorous spark that could jump across a wider air gap. Hertz had discovered what we know today to be the photoelectric effect. However, after months of characterizing the phenomenon,

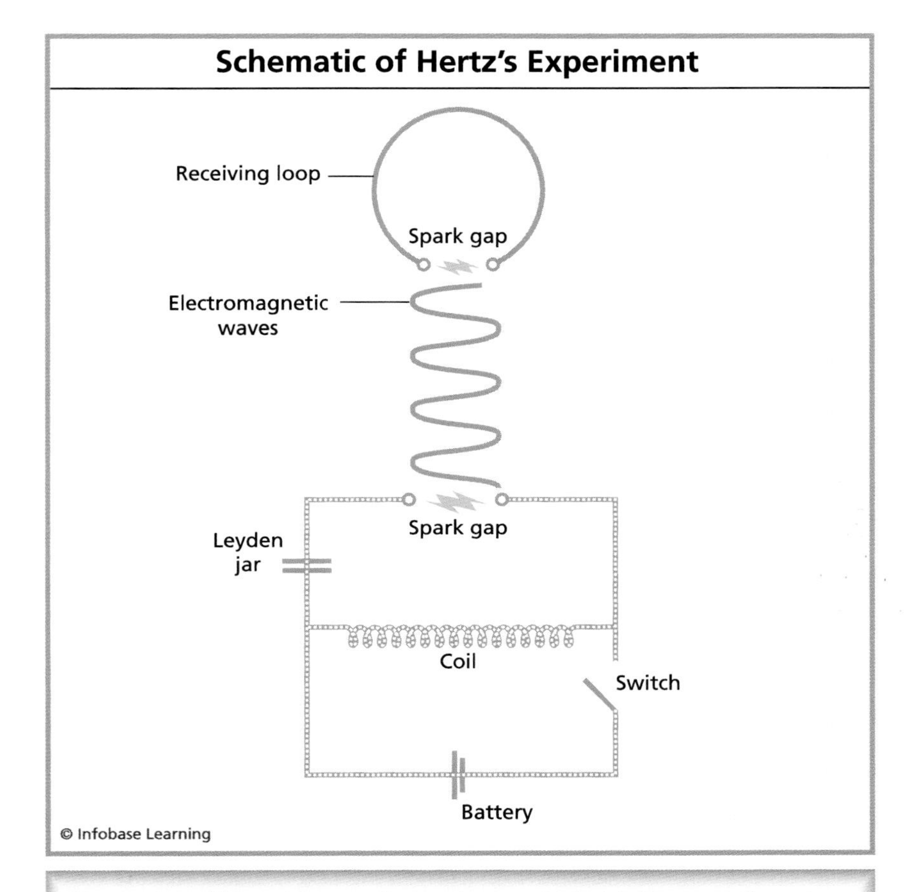

Figure 3.1 Using this experimental setup, Heinrich Hertz confirmed James Clerk Maxwell's theories about the existence of electromagnetic radiation.

he was unable to determine why light changed the maximum spark length. "I confine myself at present," he wrote, "to communicating the results obtained, without attempting any theory . . . "

While Hertz did not propose a hypothesis to explain how light amplified the spark in his receiver, his work did experimentally confirm Maxwell's theory. This confirmation pushed Hertz to forefront of physicists. Promotions and accolades followed. Unfortunately, Hertz was in poor health, and he died young, on New Year's Day in 1894, just shy of his thirty-seventh birthday.

THE MISSING PARTICLES

The main reason that Hertz attempted no hypothesis about the effect of light on the spark produced by an electrical circuit was because scientists in Hertz's day did not know the answer to the most basic question of all: What was electricity? This was a time when many scientists had not accepted the atomic hypothesis. In addition, most of those who had accepted it subscribed to the ideas of English scientist John Dalton, who proposed in 1808 that atoms were tiny, indivisible lumps of matter.

Dalton's concept of an indivisible atom, while prescient and useful, would be overturned by a series of experiments with **cathode** ray tubes (CRTs), devices that are similar to the picture tubes used in early television sets. In its simplest form, a cathode ray tube is a glass tube with most of the air pumped out. When a voltage is applied, lovely colorful patterns appear in the tube. Violet streamers stretch across empty space from the negatively charged cathode to a positively charged electrode called an **anode**. Scientists desperately wanted to know what was happening inside these tubes. As the nineteenth century wound down, evidence gradually accumulated as amateurs and scientists conducted experiment after experiment with CRTs.

They discovered that a solid object placed between the cathode and anode caused a shadow to appear at the end of the tube where the anode was located. This indicated that the discharges came from the cathode and traveled in a straight line. Because of this, the discharges were called cathode rays. One clever experimenter placed a tiny paddle wheel in the path of cathode rays. The rays turned the wheel, indicating that the so-called rays were actually particles. Yet another experiment showed that cathode rays were deflected by a magnetic field. This meant they were electrically charged particles.

The questions that remained were daunting. How big were these particles, and how much charge did they carry? And most importantly, were they atoms? The man who would answer these questions was Joseph John (J.J.) Thomson, an established scientist who held the prestigious position of Cavendish Professor of Experimental Physics at Cambridge University. As the nineteenth century wound down, he began to characterize the mysterious rays in a series of meticulous experiments.

First, he showed that cathode rays in an electric field were deflected away from the negatively charged plate. Because opposite

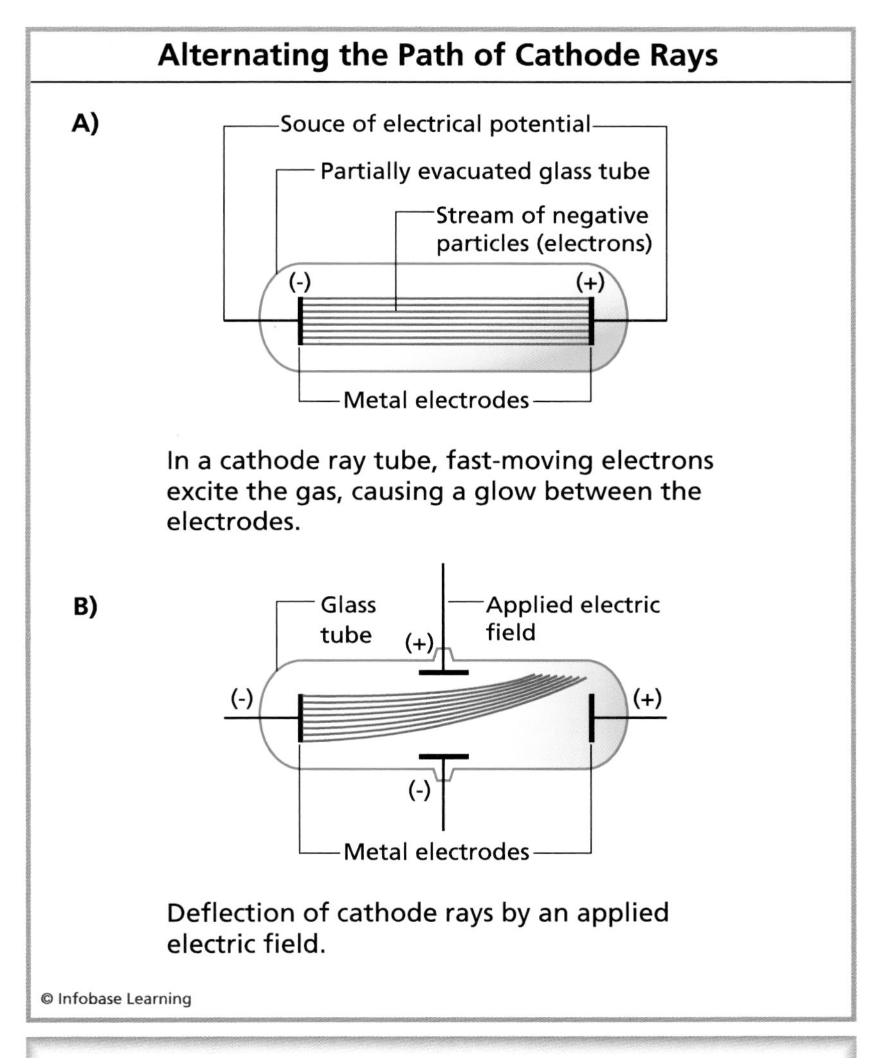

Figure 3.2 (a) The cathode ray tube was used in early experiments to characterize the electron. (b) Cathode rays are deflected by an applied electric field.

charges attract and like charges repel, Thomson concluded that the charge on the rays must be negative. Finally, by carefully measuring how much the negatively charged particles were bent by electric and magnetic fields, Thomson could calculate the ratio of the mass of the particle to its charge. To his surprise, the mass to charge ratio

(continues on page 34)

Benjamin Franklin: Master of Static Electricity

About 150 years before J.J. Thomson showed that the electrical current in a CRT was the movement of electrons through the tube, the American statesman and scientist Benjamin Franklin was experimenting with static electricity. The effects of static electricity are created by charged particles at rest. Franklin, of course, was unaware of the existence of electrons and protons, but his investigations shed a great deal of light on the nature of electricity.

After many experiments, both playful and serious, Franklin came up with a crucial concept. It was the idea that static electricity arose in matter that was positively or negatively charged. Furthermore, Franklin realized that rubbing two different substances together did not create electricity but merely moved charges from one material to the other. Rub a glass rod with a silk cloth then separate them. The glass has lost electrons to the cloth, leaving the rod positively charged while the silk has a negative charge. Now, bring the two back together, and the charges will flow back from the silk to the rod, leaving both materials electrically neutral.

Franklin also suspected that lightning was merely a dramatic manifestation of static electricity. A lightning flash, he suspected, was no more than an oversized spark, similar to the ones he could produce in his lab. Franklin's famous—and dangerous—kite experiment proved his point. The experiment showed that lightning was the same as ordinary "electrical matter," a flow of excess charges that moved from the sky to the ground, just as charges move between silk and glass. This realization led Franklin to develop the lightning rod, an invention that has saved lives and property.

Franklin was a practical man, and his contribution to electrical knowledge came long before there was much

Figure 3.3 Benjamin Franklin became interested in the principles of electricity in 1747 and published *Experiments and Observations on Electricity* in London in 1751.

theory to support his experimental work. Nevertheless, his ideas helped pave the way for the scientists who discovered and later explained the photoelectric effect.

(continued from page 31)
was one thousandth that of a hydrogen **ion**. (A hydrogen ion is a hydrogen atom that has lost its electron, giving it a positive charge.) Cathode rays either carried a huge charge or else the particle was much smaller than hydrogen, the smallest atom.

Work by other scientists showed that cathode ray particles were indeed much lighter than hydrogen atoms. This led Thomson to an astounding conclusion. Cathode rays, he announced to the world in 1897, must be a part of an atom. This was big news. All atomic theories before this one, going back to those proposed by Democritus in ancient Greece, held that the atom was indivisible. Now, here was J.J. Thomson saying it was made up of even smaller particles. These particles were soon named electrons. Furthermore, Thomson wrote, "I can see no escape from the conclusion that [cathode rays] are charges of negative electricity carried by particles of matter."

Thomson had discovered the carrier of electricity. Scientists had been probing electrical phenomena for years. Finally, they could answer the question: What is electricity? Ordinary electricity, as revealed by Thomson's work with CRTs, is the flow of electrons. (An exception is static electricity, which is produced by charged particles—electrons or **protons**—that are not in motion.) Thomson's work would prove to be one of the keys to understanding the photoelectric effect. However, before it could be completely explained, more data would be needed.

GETTING QUANTITATIVE

Working independently, Thomson and German physicist Philip Lenard began to further characterize the effects of light on electric circuits. Two years after announcing his discovery of the electron, Thomson showed that shining an ultraviolet light on a cathode made of a metal plate could induce an electric current in a CRT. The conclusion was that the light must be knocking electrons out of the cathode, causing a current to flow.

A few years later, Lenard—who had been Hertz's assistant—began measuring the maximum energy of the electrons that were knocked out of the metal plate. Surprisingly, when Lenard increased the intensity of the light by moving it closer to the plate, he found

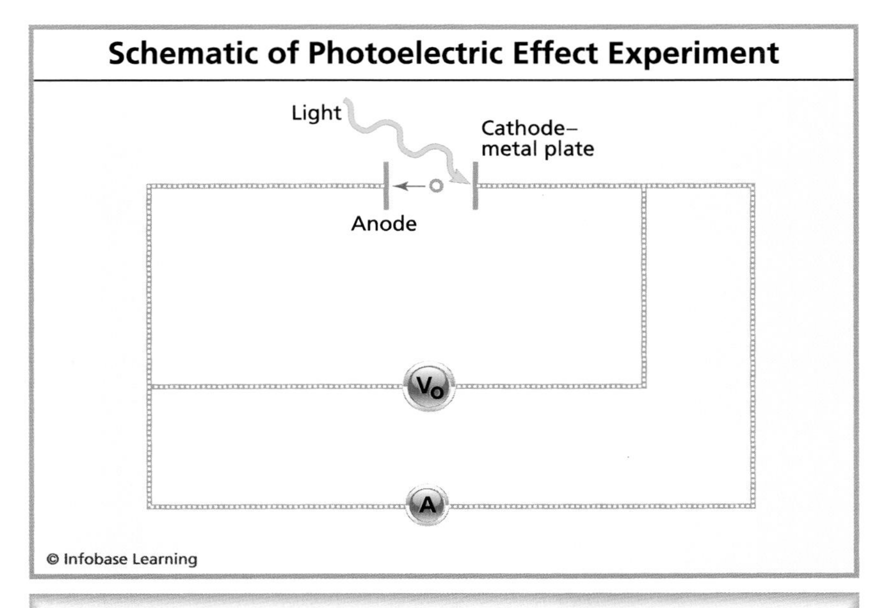

Figure 3.4 Using this circuit, one can measure both the current and the maximum kinetic energy of the ejected electrons.

that the energy of the emitted electrons did not increase. The higher-intensity light dislodged more electrons, but none of them were more energetic than the electrons produced with lower-intensity light. This was a surprise. Higher-intensity light increased the amount of energy hitting the plate. Classical physics predicted that the energy the light imparted to the electrons should also increase.

Lenard's next experiment produced equally puzzling results. When he switched to bombarding the metal plate with **monochromatic** light, he found that the maximum energies of the emitted electrons did change. Higher frequency light produced higher energy electrons. These were important, if strange, results. Both Lenard and Thomson were awarded the Nobel Prize in physics for their work on cathode rays. Lenard received his Nobel Prize in 1905 and Thomson received his a year later.

The two men had substantially advanced science's knowledge of electricity and the photoelectric effect. However, the idea of light bouncing electrons out of a metal tube meant that light was a particle with **momentum**. This finding created a huge problem for physicists because it contradicted research done a century earlier that they thought had definitively answered the question "What is light?"

THE NATURE OF LIGHT

In 1801, Englishman Thomas Young performed a series of experiments to establish the nature of light. Was it a wave or a particle? The key experiment that answered that question is known as the double-slit experiment.

In this experiment, light passes through a single slit or pinhole and continues on through a double slit. The result is a series of alternating light and dark bands called an interference pattern. This pattern occurs because of the way waves behave. When the peaks or troughs of two light waves coincide, the result is a bright band

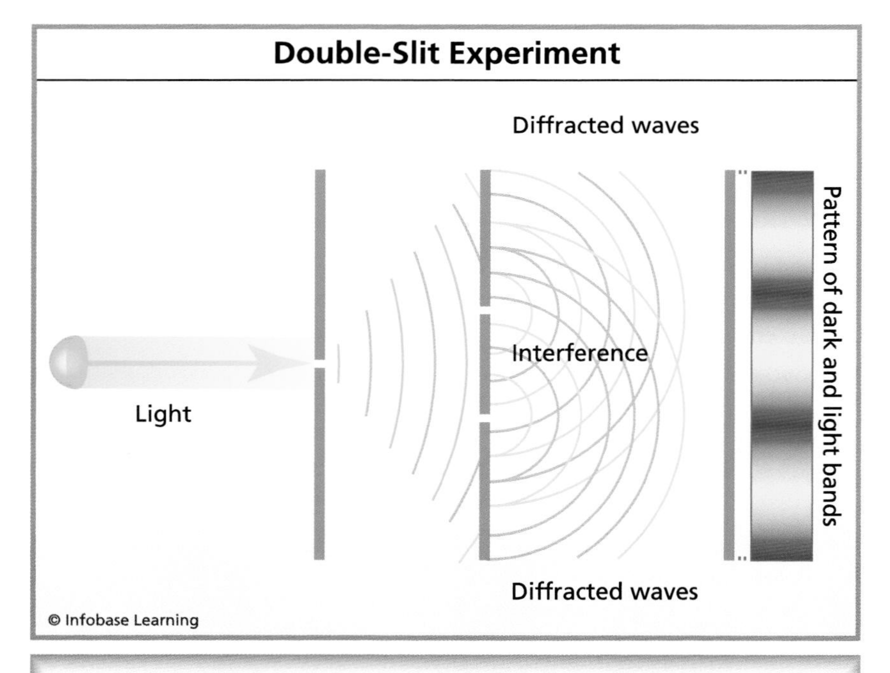

Figure 3.5 The double-slit experiment proved that light behaves as a wave. Circular waves passed through each of the holes in the screen, producing a pattern of light and dark on the viewing screen.

Figure 3.6 English scientist Thomas Young conducted experiments with light waves by experimenting with his vision in a mirror.

of light. However, when the peak of one wave coincides with the trough of another, the waves cancel each other and produce a dark band. Particles, however, do not produce peaks and troughs, so this interference pattern could not appear if light were a particle. Light, Thomas Young proved, is a wave.

However, the wave behavior of light, as established by Young, did not fit the experimental results observed by Thomson and Lenard in their work on the photoelectric effect. Higher-intensity light increased the energy falling on the metal plate. However, the emitted electrons were not more energetic. The energy of the electrons only increased when the frequency of the light increased. The discrepancy between the actual and the predicted results posed a great problem for science, and it would take a great scientist to discover the deceptively simple answer.

ENTER EINSTEIN

Albert Einstein was only 26 years old when he revolutionized physics. The year was 1905, and after years of trying to land an academic position, Einstein had given up and was working at the patent office in Bern, Switzerland. The young Einstein did not resemble the iconic, sad-eyed, frizzy-haired, older man who was to become the most famous scientist in the world. In fact, he was a handsome rebel, a Bohemian, and not at all famous. He did not get along well with authority figures, and his contempt for convention led him to father a child out of wedlock. It also made him willing to ignore accepted scientific wisdom and blaze new paths.

Living in Bern meant that the young Einstein was far from the mainstream of science, but he read the journals and kept up to date in physics. He was familiar with Max Planck's work on blackbodies and with Philip Lenard's experiments on the photoelectric effect. Unlike many other physicists, Einstein believed that Planck's equation might be more than a mathematical trick devised to solve the blackbody problem. He suspected that the equation and the discontinuity of energy it mandated might be not just a gimmick but reality.

When Einstein finally accepted Planck's equation completely, he concluded that light really did come in tiny indivisible packets of energy called quanta. Further, if that were so, he reasoned, then making the light more intense would indeed knock more electrons out of the metal. However, the energy of these packets (later named

Figure 3.7 In 1905, the year this photograph was taken, 26-year-old Albert Einstein, published three seminal papers. The first paper discussed the photoelectric effect, the second proposed what is today known as the special theory of relativity, and the third centered on a statistical interpretation of Brownian motion.

photons) from Lenard's monochromatic light source was governed by Planck's equation, which related a packet's energy to its frequency. Thus, the energy of the ejected electrons would not change, no matter how intense the light source became. More packets would strike the metal plate, but the energy of those packets would all be the same, as would the maximum energy that could be imparted to an electron. However, if the frequency of the radiation falling on the plate increased, then Planck's equation predicted that the energy of the light packets would increase, and the more energetic light packets would increase the energy of the emitted electrons.

These were exactly the results that Lenard had obtained experimentally. Einstein had explained the photoelectric effect. He had also discovered the quantum nature of light. He now knew that light was composed of particles, tiny bundles of energy called quanta.

Einstein's work solved the problem of the photoelectric effect, but it posed an even bigger question. The interference patterns from Young's double-slit experiment clearly showed that light was a wave. How could Einstein reconcile his conclusions about the particle nature of light with the wave nature of light as shown by the double-slit experiment? He wondered if light could sometimes act as a particle and other times as a wave.

Einstein was cautious about this revolutionary idea. Furthermore, he was busy sorting out another set of revolutionary ideas: the general theory of relativity. After clearing that up, though, Einstein returned to quantum mechanics and finally accepted the hard-to-accept solution: Light had a dual nature—sometimes it acted like a particle, sometimes like a wave.

As strange as this conclusion was, physics would soon get even stranger. In the decade that followed Einstein's discovery of the quantum nature of light, scientists would continue to explore the structure of the atom. And as the outlines of that structure emerged, it became obvious that atoms, too, exhibit some of the same quantum features as light.

The Bohr Atom

Cambridge University professor J.J. Thomson's postulate that electrons are subatomic particles led scientists into a new realm. They now knew that atoms were not solid, indivisible lumps of matter but contained even smaller particles, called electrons. Still, scientists wondered, how could one build an atom out of these tiny negatively charged particles? After all, atoms themselves carry no charge. To offset the negative charge of the electrons, there must be positive charges elsewhere in the atom. Seven years after discovering the electron, Thomson found the answer and completed his theory of atomic structure.

Atoms, he said, were composed of electrons distributed in a soup (or cloud) of positively charged material. The electrons were free to rotate in orbits in the soup, and their negative charges exactly offset the positively charged soup. Thomson's atom was usually pictured as a sphere with electrons scattered about in it, much as raisins are scattered about in a pudding. Because of this, it became known as the plum pudding model of the atom.

The plum pudding model was short lived. It was disproved by Ernest Rutherford, one of Thomson's best students. Rutherford was an unlikely scientist. He was born in 1871 and raised in rural New Zealand, about as far as you can get from the world's scientific centers. He became interested in science while in elementary school. He did well at it, winning scholarship after scholarship and earning degree after degree, all in physics or mathematics. At age 23, Ruther-

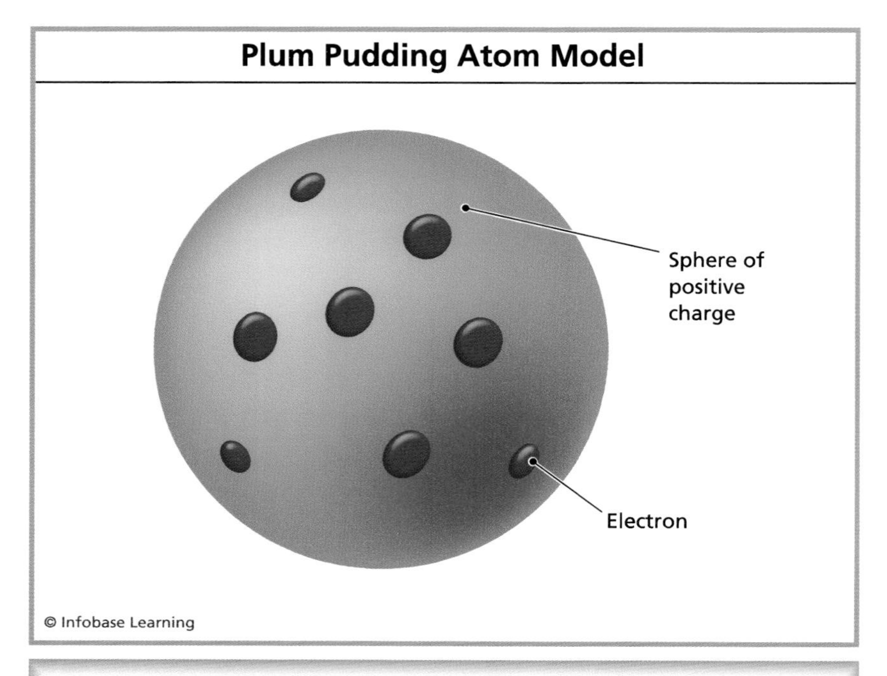

Figure 4.1 J.J. Thomson's plum pudding model of the atom consisted of electrons scattered in a positively charged soup or cloud.

ford got the job he wanted. He was awarded a fellowship to study at Cambridge. He elected to work with J.J. Thomson at the Cavendish Laboratory, the most advanced physics lab in the world.

Rutherford was a skilled experimentalist. His apparatuses were usually jury-rigged and crude, but they got the job done. His work with the particles and rays that were emitted by **radioactive elements** led him to conclude that their emissions came in two forms. With admirable simplicity, Rutherford named them after the first two letters in the Greek alphabet, alpha and beta.

After Cambridge, Rutherford accepted physics professorships at universities in Canada and, later, in Manchester, England, which was J.J. Thomson's hometown. Along the way, he continued to work with radioactive materials. Rutherford was the first person to detect radon, a new element. He also made the startling announcement that radioactivity resulted from subatomic transformations that completely changed the nature of the atoms involved. When he won the 1908 Nobel Prize in Chemistry, he quipped that although he had witnessed many radioactive transformations, none occurred as quickly as his own—from physicist to chemist. But Rutherford was only joking: He actually considered physics to be the keystone science. "All science," he relished saying, "is either physics or stamp collecting."

Rutherford accomplished most of his groundbreaking research on radioactivity before he was 40 years old. Even so, his best work was yet to come. He was full of ideas. One of those ideas was to investigate further the atomic structure proposed by J.J. Thomson. Rutherford wanted to see what happened to **alpha particles** when they were fired at a thin sheet of gold foil. Rutherford knew that alphas were much bigger than electrons and that they carried a positive charge. The plum pudding atom, composed of a positively charged soup and tiny electrons, should not change the path of a massive alpha particle. If the path of the alphas was unchanged after passing through the foil, it would add credence to the plum pudding model.

Those So-called Rays

By the early twentieth century, scientists had identified four different types of rays. Some of them would prove to be actual rays—in the form of electromagnetic radiation. Others would turn out to be particles. The table below is a catalog of the so-called rays known to Rutherford and other scientists just after the turn of the century. However, the properties given here are the modern values. The weights are based on setting the weight of a hydrogen atom as 1. The charges are based on setting the charge of a hydrogen ion as +1.

Radiation	Weight	Charge
Alpha	4	+2
Beta	5.4×10^{-4}	–1
Gamma	weightless	electrically neutral
X-rays	weightless	electrically neutral

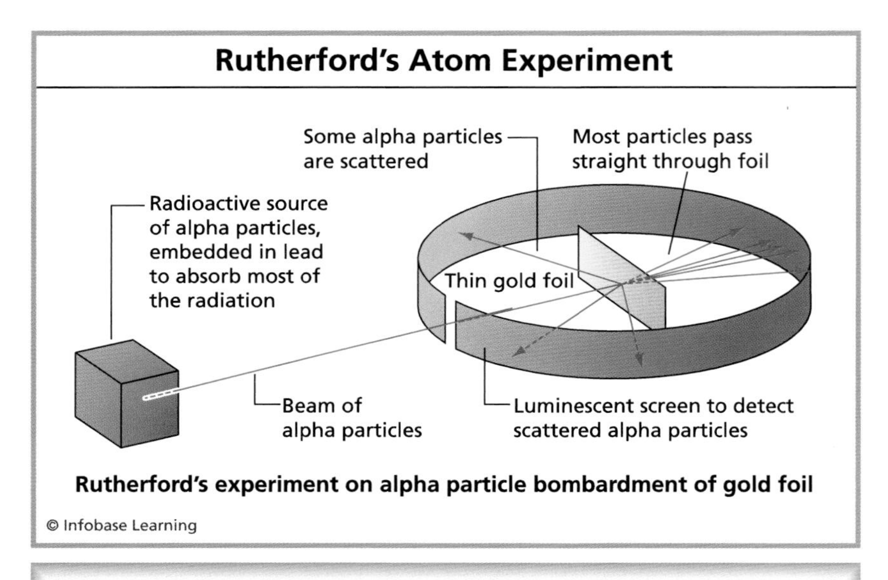

Figure 4.2 Ernest Rutherford devised this experiment to test J.J. Thomson's plum pudding model of an atom.

The apparatus that Rutherford built was simple: a source of alpha particles, a sheet of gold foil, and a detection screen that glowed briefly whenever a particle struck it .The task was tedious: count the number of glows (called scintillations) and note where on the detector they occurred. Rutherford assigned a student named Ernest Marsden to carry out this boring chore. After observing thousands of scintillations, Marsden reported that some of the particles had been deflected by large amounts, and a few had bounced directly back toward the source of the radiation. "It was," the astounded Rutherford said, "almost as incredible as if you fired a 15-inch shell at a piece of tissue paper and it came back and hit you." Clearly, gold atoms contained something more massive than electrons in a pudding. Gold was composed of something that could make an alpha particle reverse direction upon impact.

This experiment eliminated the plum pudding model as a possible structure of the atom. Still, scientists wanted to know, what did an atom look like? Rutherford figured that the only way to make alpha particles bounce backward was for the gold atoms to have a distinct area of positive charge. In a head-on collision, that charge would strongly repel the positively charged alpha particles. Also, because only a few of the bombarding particles were repelled, the atom's positive charge would have to be concentrated in a small space. Later, after more pondering, experimenting, and calculating, Rutherford announced his new structure of the atom. The atom was, he said, composed of a tiny positively charged nucleus with even tinier negatively charged electrons circling it. So small was the nucleus that if it were enlarged to the size of a marble and placed on the 50-yard line of a football stadium, then the closest electron would be found in the upper decks. Atoms, it turns out, are mostly empty space.

Rutherford published his findings on the structure of the atom in 1911. The new structure neatly fit the existing data, and it resembled a familiar arrangement from classical physics: the solar system.

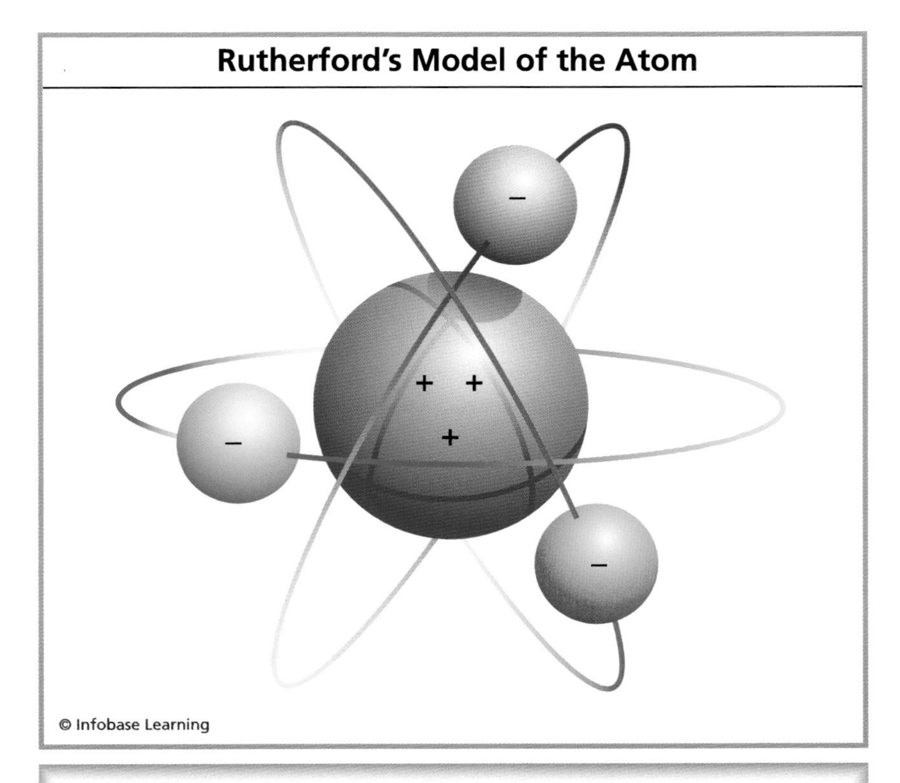

Figure 4.3 In Ernest Rutherford's model of the atom, negatively charged electrons orbited the positively charged nucleus, much like the planets orbit the Sun.

Finally, two millennia after Democritus proposed the existence of atoms, scientists had some idea of what one looked like. Unfortunately, there was a problem. Because opposite electrical charges attract one another, classical physics predicted that a negatively charged electron circling a positively charged nucleus would emit electromagnetic radiation, lose energy, and spiral down into the nucleus. According the laws of physics, Rutherford's atom could not exist.

The Discovery of Radium

Rutherford's gold-foil experiment was a first step toward a more-or-less accurate, if somewhat fuzzy, picture of the atom. This picture would catalyze the development of quantum mechanics, as scientists scrambled to understand why atoms behave as they do. However, that crucial experiment would have been impossible without a ready supply of a strong alpha-emitting element. That element was radium, and the road to its discovery began in 1895 when Wilhelm Röntgen first detected some mysterious emissions. He called them X-rays after the mathematical symbol for an unknown quantity.

Soon after Röntgen's discovery, scientists were hot on the trail of substances that emitted these unknown rays. One of those scientists was Henri Becquerel. While searching for materials that gave off the mysterious rays, he discovered that uranium salts emitted rays that could darken a photographic plate. Although the term *radioactive element* had yet to be invented, Becquerel had observed one consequence of exposure to those elements.

Shortly after Becquerel's discovery, a newly married graduate student at the Paris School of Chemistry and Physics was searching for a hot topic for a Ph.D. thesis. She decided that investigating Becquerel's rays would be suitable. With that decision, Polish-born chemist (of French citizenship) Marie Curie began checking various substances for radioactivity.

BOHR TO THE RESCUE

Niels Bohr came from one of Denmark's prominent intellectual families. His sterling background helped him receive a first-class education. After getting his college degree, he studied under J.J. Thomson in Cambridge. Then he went to work for Rutherford, one year after Rutherford had published the structure of his atom, the

Uranium, she found, was radioactive as expected. So was thorium, the next heaviest element. Surprisingly, pitchblende, a black ore rich in uranium, was four times more active than uranium itself. This led Madam Curie to suspect that a new, extremely radioactive element was present in pitchblende. She set about finding it.

Within a year, Marie and husband, Pierre Curie, had announced the discovery of two new radioactive elements. The first they named polonium, after Marie's home country of Poland. The second was called radium. However, to conclusively demonstrate their existence, the Curies needed to produce samples of the new elements that were sufficiently large to convince other scientists. Because radium was much more radioactive than polonium, they chose to isolate it first.

The concentration of radium in pitchblende is very low, so the Curies started with 10 tons (9 metric tons) of the ore to ensure they would get decent-sized sample of pure radium. To concentrate the radium, they used some elaborate separation techniques, including a series of difficult and tedious fractional crystallizations. Finally, their hard work paid off. The 10 tons of pitchblende had yielded one-tenth of a gram of pure radium chloride. Two months later, Madam Curie received her Ph.D. degree. Seven months after that, the Nobel committee awarded the Curies and Henri Becquerel the 1903 prize for physics. Their work had huge consequences, one of which was showing how to isolate the radium that Ernest Rutherford would need for his groundbreaking, gold-foil experiment.

Figure 4.4 Niels Bohr (*left*) is photographed in his laboratory in the 1940s. The Danish physicist worked as part of a team of scientists conducting the Manhattan Project, an Allied project to develop the first atomic bomb before Germany and Japan during World War II.

atom that could not—but obviously did—exist. Bohr's burning desire was to discover what held the electrons in Rutherford's atom in place. What kept the negatively charged electrons from falling into the positively charged atomic nucleus?

Bohr knew of Max Planck's work with blackbodies. What, wondered Bohr, if atoms exhibited the same quantum nature as blackbodies? What if the energies of the electrons in an atom were not continuous?

After a year with Rutherford, Bohr returned home to Copenhagen. However, progress toward a new structure of the atom was slow until he began to study the spectrum of hydrogen. When hydrogen atoms are excited by an electrical discharge, they emit radiation. The emissions appear as sharp lines of specific wavelengths. After studying the emission lines, Bohr proposed a new structure for the hydrogen atom.

Like Rutherford, he pictured the atom as having a tiny nucleus with an electron circling it like a planet orbiting the sun. However, Bohr postulated that each electron can only have certain energies. Consider a hydrogen atom with one electron and two possible energy levels, designated as n = 1 and n = 2 (Figure 4.5). (Hydrogen actually has more than two energy levels, but only two will be considered in this example.) An electron can jump from a lower energy level to a higher one by absorbing energy from a photon; or, it can go from a higher one to a lower one by ejecting a photon. However, no intermediate energy levels exist. The atom is either in one state or the other and changes instantaneously between the two.

By applying Planck's ideas about quantized oscillators to the atomic structure, Bohr solved the impossible-atom problem. The energy of an electron in an atom was fixed. It could go from one energy state to another, but it could not emit a continuous stream of radiation and spiral into the nucleus. Quantum rules prohibit it.

With this model, one could use Planck's equation to calculate the energy difference between orbits of an electron in a hydrogen atom. In the example of a system with only two possible energy levels, the equation giving the frequency of emitted radiation as the electron goes from a higher energy state, E_2, to a lower one, E_1, would be

$$E_2 - E_1 = hf$$

where h is Planck's constant and f is the frequency of the emitted radiation.

Because hydrogen has more than two energy levels, it emits electromagnetic radiation at more than one frequency. Bohr's formulation accounted for most of hydrogen's emissions. (It would soon be extended to account for all of them.) Bohr published his new atomic structure in 1913. According to Albert Einstein, "It is one of the greatest discoveries."

SPECTROSCOPY

Much of the data that supported Bohr's model of the hydrogen atom came from a branch of science called spectroscopy.

When an atom in its **ground state** absorbs a photon, an electron is promoted to a more energetic state. When the electron drops to a

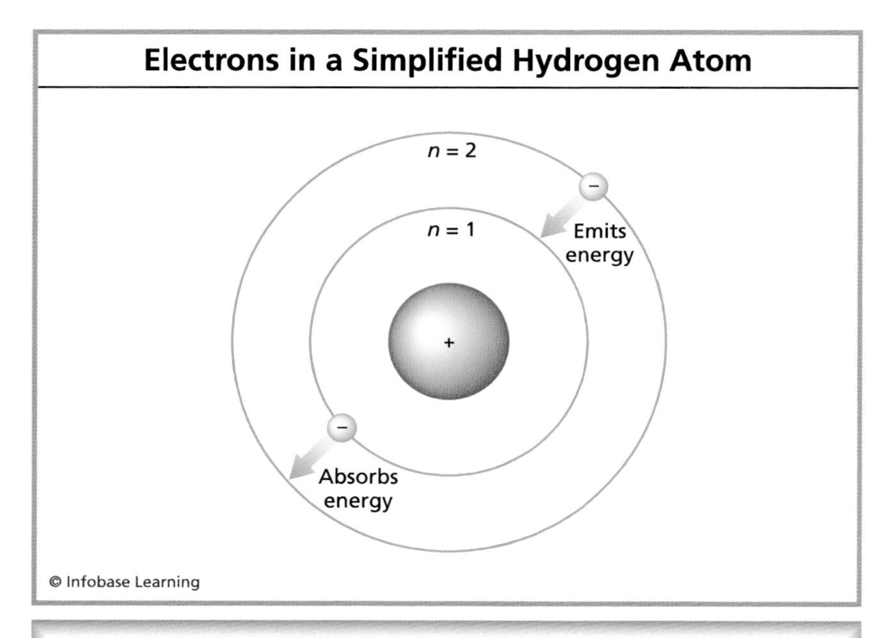

Figure 4.5 By ejecting a photon (emitting energy), an electron can move from a higher energy level to a lower one. Conversely, by absorbing a photon, an electron can jump from a lower energy level to a higher one.

lower energy state, it emits energy, usually in the form of a photon. Spectroscopy is the branch of science that investigates that quantum of emitted or absorbed energy.

Absorption spectroscopy deals with atoms capturing a photon that bumps an electron into a higher energy state. Emission spectroscopy uses an external source of energy—heat, radiation, or an electrical current, for instance—to excite the electrons in an atom. When the electrons fall from an excited, higher-energy state to a lower energy level, they emit a photon. Spectroscopists measure the wavelength of the emitted photons. Using Planck's equation, they can calculate the energy released in the transition.

By the early twentieth century, scientists had analyzed the spectra of most elements. They knew that each element produced a characteristic emission spectrum. Because hydrogen is the simplest atom, much of the research to understand the nature of atomic spectra centered on it.

Send an electric current through a glass tube containing hydrogen at low pressure and a blue light will appear. When this light passes through a prism, four colored lines show up—red, blue-green, blue, and violet. This series of lines in the visible range were the first to be discovered in the early days of spectroscopy. Swiss physicist Johann Balmer developed an equation in 1885 that enabled him to calculate the wavelengths of the lines. His equation also predicted the existence of other spectral lines for hydrogen, including one near the edge of the visible spectrum; this line was detected soon afterward. To honor his contributions to spectroscopy, this series of visible spectral lines is called the Balmer series.

The discovery of two other series of emission lines of hydrogen came later. They are named for their discoverers: the Lyman series, which is in the ultraviolet range, and the Paschen series, which is in the infrared region. Although formulas were devised to calculate the spectral lines, the physics behind the formula was not understood until Niels Bohr proposed his quantized atom. Suddenly, the emission spectrum of hydrogen made sense. Each line represented the energy released when an excited electron went from a higher-energy quantum state to a lower one.

Over time, scientists sorted out the electron transitions that produce every line in the spectrum of hydrogen, as shown in Figure 4.5. This was the experimental skeleton on which Bohr hung his new atomic model. The high-energy Lyman series comes from transitions from higher-energy states to the ground state of hydrogen, n = 1. The less energetic Balmer series in the visible region involves electrons dropping to the n = 2 energy level. And the low-energy Paschen series in the infrared region comes from electrons going into the n = 3 energy level. The reason more energy is emitted when electrons transition to the n = 1 level is because the difference in energy levels increases as *n* gets smaller. Thus, a large dollop of energy is emitted when electrons transition to the n = 1 level, but less energy is emitted in transitions into higher energy levels, such as n = 2 or n = 3.

The Bohr atom explained the spectrum of hydrogen, but its basic concept was a hybrid. It featured a strange combination of classical physics (tiny negatively charged particles circling a positively charged nucleus) and quantum ideas (fixed orbits for the electrons). And extending it to atoms heavier than hydrogen would prove to be much more difficult and give far less accurate predictions.

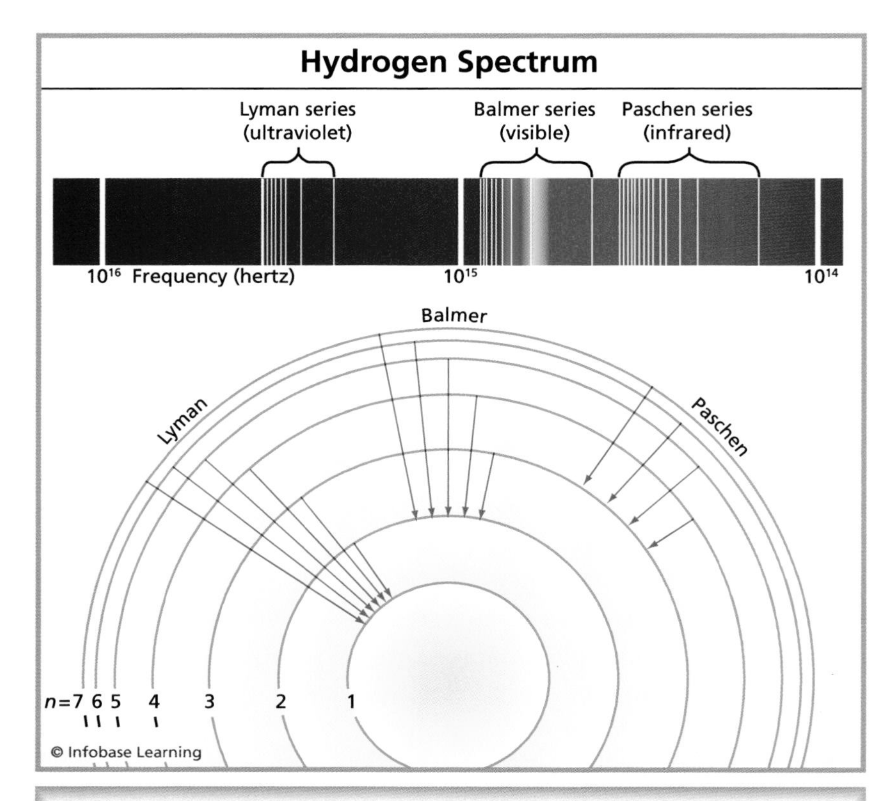

Figure 4.6 The lines in the hydrogen spectrum correspond to the emission of energy when an electron drops to a lower energy level.

BEYOND HYDROGEN

Hydrogen is the simplest atom, consisting of a nucleus of one positively charged proton with one negatively charged electron circling it. Bohr knew his theory of the atom had to be extended to other elements. To account for the properties of other atoms, Bohr borrowed a concept originally introduced by J.J. Thomson. The idea was that electrons in atoms occupy shells surrounding the nucleus. In the shell model, an atom can be thought of as an onion, with each layer of onion representing one shell. Bohr extended Thomson's idea by assigning specific energies to each shell. Following the nomenclature he established for the orbits of hydrogen, he labeled the lowest energy shell as n = 1, the next higher as n = 2, and so on.

Using this concept, Bohr could build imaginary atoms—electron by electron. After hydrogen came helium with a nuclear charge of +2. Helium is a very stable element, reluctant to lose or gain electrons. So, Bohr concluded that 2 electrons completely filled the first energy shell in an atom. Additional electrons would have to go into another shell. Bohr then determined that it took 8 electrons to fill the next energy shell. Drawing on his knowledge of the periodic nature of the elements, spectroscopic data, and the natural intuitive genius that marked his entire career, Bohr extended his atomic theory from hydrogen to all the elements.

The Bohr atom went a long way toward explaining the nature of atoms. Still, there were problems. Scientists could calculate the emission spectrum of hydrogen using the Bohr model, but they could not account for the spectra of heavier atoms. However, the biggest problem with the Bohr atom lay in its arbitrary nature. The model accurately predicted some experimental results, but it had no solid theoretical footing.

What determines the energy levels of the electrons in a shell? Why are 2 electrons enough to fill the first energy shell in an atom, while 8 electrons are required for the next one? While some scientists struggled to understand the laws that governed Bohr's atom, others were working on a different problem. That problem, when solved, would lead to a new structure that would replace Bohr's solar system model of the atom. The problem was one that Einstein had wrestled with earlier: Was light a wave or a particle?

5

Building a Sound Foundation

After deep consideration, Einstein concluded that light could act as either wave (as in Young's double-slit experiment) or particle (as in the photoelectric effect). This was a difficult pill for older, classically trained physicists to swallow, but younger scientists had less trouble with it. Surprisingly, though, it was not one of the young Turks who extended Einstein's idea about the dual nature of light but, rather, a middle-aged man. In his 1924 Ph.D. thesis, a 33-year-old Frenchman named Louis de Broglie offered a far-reaching hypothesis: If light waves could act like particles, then maybe particles could act like waves.

At the time de Broglie published his hypothesis, the **quantum theory** was a mess. It explained some things: blackbody radiation, the photoelectric effect, the impossible atom, and, thanks to Niels Bohr, most of the emission spectrum of hydrogen. However, each line of thought seemed unconnected and many questions remained unanswered. How were Max Planck's oscillators related to Bohr's atoms? Why do the energy shells of atoms fill as they do? Why is the Bohr atom unable to predict much of the spectra of atoms that are heavier than hydrogen? How do classical physics and quantum theory fit together? De Broglie's thesis would be the first step toward answering these questions and toward firming up the foundations of quantum mechanics.

THE PRINCE

Scientists come from every social and economic background. Michael Faraday (1791–1867)—the inventor of, among other things, the electric motor—began his working life as a dirt-poor bookbinder's apprentice. Henry Cavendish (1731-1810)—the first man to determine the density and mass of Earth—was reputed to be the richest man in England. A few are aristocrats and have begun their lives in a titled family. One such scientist was Prince Louis-Victor de Broglie, the son of the Duc de Broglie.

Louis's family moved to Paris when he was still young. During World War I, he served in the army. Thanks to his background in science (and a brother with political pull), he managed to stay out of the trenches and was assigned to a radiotelegraphy unit located at the Eiffel Tower. Despite his aristocratic background, de Broglie did not play the role of a rich Parisian playboy. He was a hard worker and a dedicated scientist, who, according to his biographer, "led a rather withdrawn life and never married."

De Broglie's big contribution to quantum theory was to extend Einstein's idea about the dual nature of light to electrons and beyond. By the time he was choosing his Ph.D. topic, the American physicist Arthur Compton had shown that X-rays acted very much like ordinary light, exhibiting properties of both waves and particles (or photons, as they would soon be named). No one knows exactly how de Broglie came up with his big idea, but his reasoning must have proceeded something like this: The work of Einstein and Compton both indicate that electromagnetic radiation can sometimes act as a particle and, at other times, act as a wave. Then, De Broglie might have wondered, why can't electrons, which are very small particles, also sometimes act as waves?

This kind of idle speculation is common among scientists. Most of the time, however, such speculations go nowhere. The idea turns out to be a bad one or is not followed up due to the publicity of other work. No reputable journal would accept a paper based only on unsupported conjecture. Fortunately, Louis de Broglie did not stop with just an idea. Beginning with Einstein's equations from relativity and the photoelectric effect, de Broglie used some high-powered mathematics to derive the following fundamental equation that links mass to wavelength:

$$\lambda = h/p$$

In this equation, λ is the wavelength of a particle, p is its momentum, and h is Planck's constant. Momentum equals the mass of a particle times its velocity, so de Broglie's equation can be rewritten as

$$\lambda = h/mv$$

where m is the particle's mass and v is its velocity. Notice that this equation places no restrictions on mass. De Broglie's equation means that every object from electrons to elephants has a wave associated with it. This was a major shift in the thinking of the day. Only a few scientists had accepted Einstein's conclusion that light could act as a wave or as a particle. Now, here was Prince Louis-Victor de Broglie proclaiming that all matter had waves as-

Where Are the Waves?

Scientists have now established that particles much bigger than electrons—neutrons, for instance—also exhibit wave-like properties. However, de Broglie said that all matter has wave properties, including familiar objects such as golf balls, chairs, and your kid sister. Is this really true: do golf balls actually travel in sine waves after being smacked? As noted earlier, speculation is cheap in science. So, let's do the arithmetic. The starting point is de Broglie's famous equation:

$$\lambda = h/mv$$

A typical golf ball weighs about 1.6 ounces or 46 grams (g). When struck by a hard hitter using a driver, the ball leaves the tee at about 150 miles per hour or about 67 meters per second (m/s). Now, put all the variables in standard SI units:

$h = 6.6 \times 10^{-34}$ joule seconds (J•s)
$m = 0.046$ kilograms (kg)
$v = 67$ m/s

sociated with it. Einstein understood immediately the significance of de Broglie's work. "He has lifted a corner of the great veil," he wrote to a friend.

Before searching for the waves associated with ordinary objects, recall that the numerator in the equation is *h*, Planck's constant, which is a very small number. Therefore, more massive objects (such as elephants) will have a very short wavelength associated with them. To test de Broglie's theory by detecting the wavelike character of matter, scientists knew they had to experiment with particles with very low masses. The particle with the lowest mass known at the time was the electron.

Three years after de Broglie published his thesis, physicists demonstrated the wave character of electrons by producing the

Calculating for wavelength gives

$$\lambda = \frac{6.6 \times 10^{-34}\ \text{J} \bullet \text{s}}{(0.046)(67)\ \text{kg} \bullet \text{m/s}}$$

$$\lambda = \frac{6.6 \times 10^{-34}}{3.08} = 2.14 \times 10^{-34}\ \text{m}$$

In decimal form, the length of the wave associated with a golf ball leaving the tee is

0.00000000000000000000000000000000021 m

So, golf-ball waves really do exist, but the length of those waves is extremely short, more than 20 orders of magnitude smaller than an atom. Their small size makes them impossible to detect or affect the flight of the ball. The curves that the ball sometimes takes on the way from tee to green has nothing to do with waves. Hooks and slices are not quantum effects.

diffraction pattern that he predicted. Researchers Clinton Davisson and his junior partner Lester Germer, who were working in the United States, and George Thomson, who was working independently in Great Britain, made this discovery. Not only did both experiments show that electrons could act like waves, they also showed that electrons obeyed de Broglie's equation that related wavelength to momentum. Scientists now knew that matter could behave like a wave.

The Nobel Committee, sometimes considered to be slow in recognizing scientific achievement, was quick to reward this research. De Broglie was awarded the Nobel Prize in physics in 1929. Davisson and Thomson got theirs in 1937. Probably no other fact brings home the dual nature of matter better than this award. George Thomson was J.J. Thomson's son. So, as J.J. Thomson won his Nobel Prize for demonstrating that electrons were particles, his son later won for demonstrating they were waves.

NIELS BOHR'S INSTITUTE

After de Broglie proposed his wave hypothesis, the pieces were in place for a new atomic structure to emerge. Bohr's picture of the atom as a dense, positively charged nucleus circled by tiny, negatively charged electrons in fixed orbits would soon be superseded by the modern picture of the atom. Much of the upcoming revolution would take place in Bohr's hometown of Copenhagen, Denmark, at a newly minted center of scientific research called the Institute for Theoretical Physics.

Today, the institute is a sizable operation with 12 science centers, 10 research groups, over 200 employees, and many aspiring physics students. Back in its early days, though, its heart was one small building located near the soccer fields in a park in Copenhagen. The building was constructed in 1921 to enable Bohr and his students to pursue their work in quantum theory. The institute became a magnet for up-and-coming young physicists. One of the first to show up was Werner Heisenberg, who would later win a Nobel Prize for his role in the development of quantum mechanics. Soon afterward came George Gamow, the fun-loving Russian physicist who sorted out the nuclear reactions that power the stars. Erwin Schrödinger, who would also win a Nobel Prize in physics, stopped by to lecture on occasion. Wolfgang Pauli, who would have to wait until 1945 for his Nobel Prize, was there, too.

It was an informal place. Ping-Pong and cowboy movies were favorite relaxations. However, the men who gathered there were serious scientists who were working to bring about the biggest shift in scientific thinking since Newton introduced his laws of motion. The atmosphere was collegial but irreverent. Pauli was one of the brashest of the bunch. He had demonstrated this a few years earlier. After Albert Einstein delivered a lecture at his university, Pauli—only a teenager at the time—stood up and said, "You know, what Mr. Einstein said is not so stupid. . . ."

Great scientists came and went at the institute, but one thing remained constant. That, of course, was Niels Bohr himself. He was kindly, brilliant, and well connected. Thus, it was only fitting that in 1965, three years after his death and on his birthday, the Institute for Theoretical Physics was renamed the Niels Bohr Institute.

HEISENBERG LEADS THE WAY

Just as scientists come from all walks of life and social classes, they also come in various physical forms. However, scientists spend long hours at their desks or in the lab. Consequently, the stereotype is that many of them tend to occupy themselves with less-strenuous hobbies that do not lead to an athletic body. Two of the most prominent participants in the Institute for Theoretical Physics were among those that did not fit this image. Bohr, for one, was a strong, energetic skier. He loved soccer and played on Denmark's amateur team. In 1924, he was joined by another good athlete, the young Werner Heisenberg.

Heisenberg was born in 1901 in Würzburg, Germany. He loved the outdoors, enjoying long camping and hiking trips. He was, associates said, "a 'radiant' young man, well endowed in body and brain, and extremely self confident." "The world was his oyster," according to one acquaintance.

It certainly seemed that way. Heisenberg led a charmed academic life, getting his Ph.D. in theoretical physics in 1923, at the age of 21. He had fallen under Bohr's sway two years earlier and visited him often in Copenhagen. Although Heisenberg had an enormous respect for Bohr, he harbored doubts about Bohr's formulation of atomic structure. In the spring of 1925, Heisenberg began to think hard about the quantum nature of atoms. He first stripped the problem down to its essentials. He asked himself: What do we actually know about atoms?

Some facts had been established. The basic constituents of the atom—a positively charged nucleus and negatively charged electron—were known. Spectroscopic experiments indicated that the electrons could move from one energy state to another by absorbing or emitting electromagnetic radiation. All of these observations fit in with Bohr's atomic structure. However, nowhere in all of this experimental knowledge was the evidence that electrons circled the nucleus in neat, well-defined orbits. Although those fixed orbits were the key to Bohr's atomic structure, Heisenberg saw no evidence that they actually existed.

As he worked, Heisenberg discovered that he could not represent the quantum states of an atom with ordinary numbers. He needed arrays, in which numbers or variables are arranged in rows and columns. As Heisenberg worked with the arrays, he discovered that the results he obtained upon multiplying two arrays depended on the order in which one does the multiplying. In mathematical terms, such behavior means that the two arrays that are being multiplied do not **commute**; that is, the result of multiplying two arrays, let's call them A and B, depends on the order in which they are multiplied. Thus,

$$A \times B \neq B \times A$$

Despite this peculiarity of the mathematics, Heisenberg suspected that he was on the right path to an improved quantum theory, one with a firm theoretical foundation. However, he was tired after the marathon session he put himself through to produce the new theory. So, he hurriedly wrote up his notes and left them with his boss, Max Born, who was chairman of the physics department at the University of Göttingen in Germany.

Born quickly realized that the arrays and laborious mathematics constructed by Heisenberg were already known. The arrays were **matrices** and the rules for manipulating them, including multiplication, had been worked out years earlier. For assistance in converting

Figure 5.1 Young, confident Werner Heisenberg poses in 1927, the year he published his uncertainty principle.

Heisenberg's calculations to the language of matrix algebra, Born called on his young assistant, Pascual Jordan, for help. During the summer, Born and Jordan worked on Heisenberg's manuscript. Later, Heisenberg joined them by mail from Copenhagen where he was consulting with Bohr. At first, Heisenberg was baffled. ". . . [b]ut I do not even know what a matrix is," he told Bohr. Still, he communicated regularly with Born and Jordan by mail and quickly picked up the elements of matrix mechanics.

The result of this collaboration was the first coherent mathematical framework for the quantum theory. It appeared in a paper in September 1925. For the first time, Heisenberg felt comfortable with the foundations of quantum theory. Further, as he had predicted earlier, the new theory dismantled Bohr's atom. ". . . [i]t seems," Heisenberg said, "as if the electrons will no more move on orbits."

Heisenberg's new mathematics were an immediate success. Wolfgang Pauli, the teenaged whiz who thought Einstein "was not so stupid," used Heisenberg's matrices to calculate the spectrum of hydrogen. The results coincided perfectly with experiment. All told, Heisenberg's new method of doing quantum mechanics was a raging success. It was also about to be replaced—not by a more accurate model, but by one that was much easier for scientists to understand and use.

SCHRÖDINGER'S WAVES

"No mathematician should ever allow himself to forget that mathematics, more than any other art or science, is a young man's game," wrote G.H. Hardy, in his book, *A Mathematician's Apology*. The

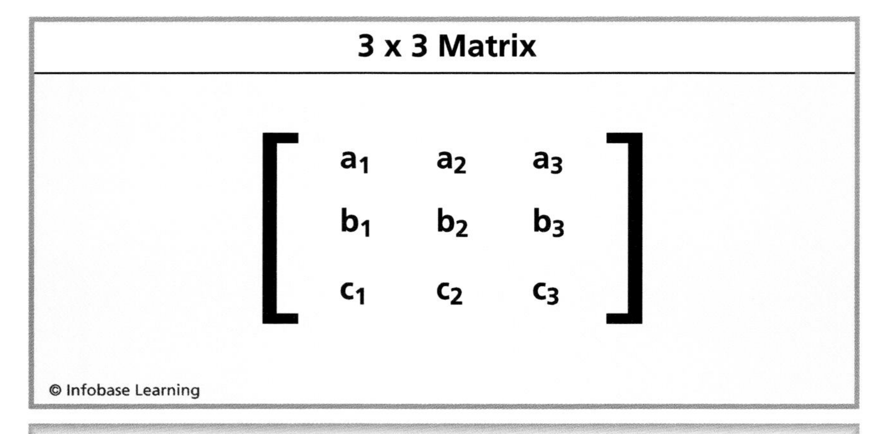

Figure 5.2 This is an example of a 3x3 matrix.

same limitation is said to apply to theoretical physicists, too. High-powered mental activities, such as mathematics and physics, are a young man's game. If you have not made your big breakthrough before age 30, it just might not happen. After all, Einstein was in his mid-20s when he revolutionized physics in 1905.

There are exceptions, of course. De Broglie was one of them. Another one was Erwin Schrödinger, a middle-aged Austrian physicist who would, independently of Heisenberg, produce a new method of doing quantum mechanics. His approach would displace matrix mechanics as the preferred method for quantum calculations.

During a Christmas break from his teaching duties at the University of Zurich, Switzerland, Schrödinger fled an unhappy home life to spend time with a girlfriend in a remote Alpine resort. Maybe it was his girlfriend or the cold mountain air, but whatever it was, Schrödinger was inspired to take de Broglie's work a step further. De Broglie said that all matter had a wave component, and Schrödinger aimed to expand that idea into a wave theory of atomic structure.

Schrödinger knew that under some conditions, solutions to wave equations exist only for discrete, noncontinuous values of energy that are called **eigenvalues**. Because eigenvalues are always discrete numbers, they could be considered to be "quantized." Might there be a link between waves and quanta? Schrödinger thought so, and, in January 1926, he submitted a paper entitled "Quantization as an Eigenvalue Problem." With this paper, wave mechanics was born.

The key idea is that each measurable parameter in a physical system has a quantum mechanical operator associated with it. One such operator, the **Hamiltonian**, contains the operations associated with a particle's energy. Operating on the wave function of a particle produces specific energy eigenvalues as shown here:

$$H\Psi = E\Psi$$

This is a highly simplified form of the Schrödinger wave equation. In it, Ψ is the wave function of a particle, E is the energy of the system and an eigenvalue of the wave function, and H is the Hamiltonian operator. Behind this seemingly simple equation lies a vast mathematical thicket. To use the equation to calculate the energy levels in an atom or molecule is quite complicated, but to those

Figure 5.3 In 1926, Erwin Schrödinger published a paper called "*Quantisierung als Eigenwertproblem*" ("Quantization as an Eigenvalue Problem"), which introduced his wave theory of the atom.

who can follow the math, the equation and the ideas embedded in it are a thing of beauty. J. Robert Oppenheimer, the distinguished American physicist and leader of the Manhattan Project (the project that produced the first atomic bomb), summed up his feelings about

More About Matrices

A matrix is a set of quantities laid out in a rectangular array of vertical columns and horizontal rows. Matrices may be added to, subtracted from, and multiplied by one another. Other, more complicated, mathematical manipulations are also possible. The difficulties with Heisenberg's matrix formulation of quantum mechanics becomes apparent when one multiplies the two of the simplest matrices, which we will call A and B.

$$A = \begin{pmatrix} a & b \\ c & d \end{pmatrix} \qquad B = \begin{pmatrix} e & f \\ g & h \end{pmatrix}$$

The rules of matrix algebra specify how to multiply the two matrices.

$$AB = \begin{pmatrix} (ae + bg) & (af + bh) \\ (ce + dg) & (cf + dh) \end{pmatrix}$$

Now, plugging in some numbers.

$$A = \begin{pmatrix} 0 & 1 \\ 0 & 0 \end{pmatrix} \qquad B = \begin{pmatrix} 0 & 0 \\ 1 & 0 \end{pmatrix}$$

and

$$AB = \begin{pmatrix} (0 \cdot 0 + 1 \cdot 1) & (0 \cdot 0 + 1 \cdot 0) \\ (0 \cdot 0 + 0 \cdot 1) & (0 \cdot 0 + 0 \cdot 0) \end{pmatrix} = \begin{pmatrix} 1 & 0 \\ 0 & 0 \end{pmatrix}$$

Is matrix algebra really noncommutative? The result when A is multiplied by B is shown above. So, what happens when B is multiplied by A?

$$BA = \begin{pmatrix} (0 \cdot 0 + 0 \cdot 1) & (0 \cdot 0 + 0 \cdot 0) \\ (0 \cdot 0 + 1.1) & (0 \cdot 0 + 1 \cdot 0) \end{pmatrix} = \begin{pmatrix} 0 & 0 \\ 1 & 0 \end{pmatrix}$$

$$\begin{pmatrix} 1 & 0 \\ 0 & 0 \end{pmatrix} \neq \begin{pmatrix} 0 & 0 \\ 1 & 0 \end{pmatrix}$$

(continues)

(continued)

Clearly, matrix multiplication does not commute. Clearly, too, the manipulation of matrices is tedious and difficult. Heisenberg, Born, and Jordan pulled off an amazing feat when they formulated quantum mechanics using matrices. Matrix mechanics is still used today in a few applications, but it is no wonder that most scientists prefer the easier-to-understand and easier-to-visualize wave mechanics developed by Erwin Schrödinger.

Schrödinger's equation: "Here is this quite beautiful theory, perhaps one of the most perfect, most accurate, and most lovely man has discovered."

The next chapter will show how scientists have used this "quite beautiful" wave theory to explain the properties of atoms, which is the key to the science of chemistry.

Chemistry Revealed

Remember Ernest Rutherford's comment implying that the only real science is physics? The other sciences were just playing at the margins, "stamp collecting," as he put it. It is most likely that few chemists agreed with him. Chemistry was, by then, a long-established, scientific discipline, one that produced the things needed for the good life—dyes, medicines, and fertilizers, to name a few.

Much of the understanding of how to manipulate atoms to produce these goods came after centuries of experiments. When you add a little of this to a little of that, you get something that people find useful. Still, chemistry was not just a hit-or-miss endeavor. John Dalton, Antoine-Laurent Lavoisier, Joseph Priestley, and many other chemists had introduced unifying concepts to the science. Dmitri Mendeleev brought even more order to chemistry when he developed the periodic table of the elements. The table, with its orderly rows and columns, had prognostic power. It enabled Mendeleev to predict the existence of some elements before they were discovered.

Still, Rutherford may have had a point. The properties of many atoms (and molecules) were known, but why they exhibited those properties was unclear. Not only that, as Rutherford was no doubt aware, possibly the biggest step toward understanding the properties of atoms came from a physicist: J.J. Thomson, the man who discovered the electron.

Today, scientists know that the chemical properties of an element are largely determined by the number and energy of its electrons. This chapter will explore how electrons are arranged in atoms and how the work on quantum theory led to the discovery of those arrangements.

BACK TO BOHR

Let's begin with Niels Bohr's atomic structure. In his early quantum model, electrons were particles circling a nucleus in shells that determined their energies. Bohr used the letter *n* to designate energy shells with n = 1 being the shell with the lowest energy, n = 2 the next highest, and so on. Helium, Bohr knew, has two electrons. It is a noble gas, a very stable atom that refuses to gain or lose electrons under most conditions. Bohr concluded that two electrons filled the lowest energy shell.

Electrons in atoms that are heavier than helium, Bohr postulated, must go into higher energy shells. Thus, lithium, with an atomic number of 3, has two electrons in the n = 1 energy shell. Because that shell is filled, the third electron must go into a new energy shell with n = 2.

The number of electrons required to completely fill an atom's energy shells was worked out by extending Bohr's ideas about helium to the other **noble gases**. All of these gases are very stable. They do not react easily with other substances. This means they do not gain or lose electrons readily. By 1916, Bohr and others were suggesting that these gases must have energy shells that are filled and, therefore, they can take no more electrons. Scientists now know that Bohr and his colleagues were right. All of the lowest energy shells of every noble gas are filled (Table 6.1).

Drawing on his knowledge of chemistry, Bohr extended his ideas about the arrangement of electrons in atoms to all the elements. The resulting electron configurations were a logical outgrowth of his quantized atomic structure. In many ways, the Bohr atom was a remarkable success. However, over time, problems arose. Higher resolution spectroscopes revealed new lines in the spectra of many elements. Bohr's atomic model could not explain this so-called fine structure. Nor could it completely explain the spectra of atoms

Table 6.1: Electronic Configurations of Noble Gas Atoms						
Element	Atomic Number (Z)	Number of Electrons in Energy Shell n=1	n=2	n=3	n=4	n=5
Helium	2	2				
Neon	10	2	8			
Argon	18	2	8	8		
Krypton	36	2	8	18	8	
Xenon	54	2	8	18	18	8
Radon	86	2	8	18	32	18

larger than hydrogen. However, the biggest problem with the Bohr model was its empirical nature. What was magic about the noble gases' electron shells? Why did two electrons fill n = 1 shells? Why did it take eight when n = 2? These questions were answered when Heisenberg published his matrix theory and Schrödinger developed wave mechanics.

Both theories offered a new way of looking at an atom. Gone were the certainties of Bohr's atom, replaced by de Broglie's strange wave-particle duality. Furthermore, as Max Born showed in 1926, the wave equation led to an atom governed by probabilities. No longer could one say an electron is here or there. An electron in an atom could be anywhere, although some locations are more likely than others.

Electrons exist in **orbitals**, but in wave mechanics, those orbitals no longer represent particles circling a nucleus. In the new atom, an orbital is defined as an area where there is a high probability of finding an electron. Surprisingly, this shadowy character of electrons led to a clearer understanding of how atoms are built and why they act as they do.

QUANTUM NUMBERS

Today, scientists know that the energy and behavior of electrons in an atom are determined by a set of four quantum numbers. The wave equation, shown in the previous chapter, is a simplified form of the three-dimensional, high-powered equations needed for real atoms. Nevertheless, the math can be reduced to three equations. The first three **quantum numbers** and the values they can assume arise from the solutions to those equations. The first of these is the **principal quantum number**. Following Bohr's convention, it is designated by the letter *n*. The results from Schrödinger's equations specify that it must be a positive whole number, such as 1, 2, or 3.

Quantum numbers can be considered to be approximately equivalent to physical features in the atom that was proposed by Bohr. The principal quantum number corresponds to one of Bohr's energy shells. It is roughly related to the average distance of the high probability part of an electron's orbital from the nucleus. Electrons with larger values of *n* are more energetic and are located farther from the nucleus.

The second quantum number is called the **angular momentum quantum number**. It is designated by the letter ℓ and can be thought of as representing a **subshell** within a principal energy shell. This quantum number governs the **angular momentum** of the electrons and determines the shape of the orbitals in an atom. Remember that orbitals no longer represent the path of an electron around a nucleus. In the wave-equation world, orbitals merely indicate where an electron is likely to be. The angular momentum quantum number can be any positive integer between 0 and $n-1$. So, up to three orbitals could be present in an energy shell with a principal quantum number of 3 ($n = 3$). They would have angular momentum quantum numbers of $\ell = 0$, $\ell = 1$, and $\ell = 2$.

Chemists designate angular momentum quantum numbers by the letters given in Table 6.2. The convention for naming orbitals begins with the number of the principal energy shell followed by the letter assigned to the angular momentum quantum number. A hydrogen electron in its ground state (or lowest energy level) would occupy a 1*s* orbital, where the 1 specifies the principal quantum number and the *s* denotes a subshell with an angular momentum quantum number of 0. If the electron jumped to the next higher energy level, its orbital would be in the 2*s* subshell. Similarly, the lowest energy *p* orbital would be 2*p*. Table 6.3 shows the allowed subshells for the first four principal energy shells of an atom.

The **magnetic quantum number**, usually designated as m_ℓ comes from the solution to the third wave equation. Allowable values of this quantum number range from - ℓ to + ℓ. A summary of the possible values of all three quantum numbers allowed by the wave equations for the first four energy shells is shown in Table 6.4. The magnetic quantum number specifies how the orbitals in the *s*, *p*, *d*, and *f* subshells are oriented in space. The shapes of the first three s orbitals are spherical, with the lower-energy orbitals nested inside the higher-energy orbitals (Figure 6.1).

Table 6.2 Letter Designation of Subshells

Value of ℓ (Subshell)	Letter
0	*s*
1	*p*
2	*d*
3	*f*
4	*g*

Table 6.3 Allowable Subshells in the Principal Energy Shells (*n*) of an Atom

n	Subshell Integer (ℓ)	Subshell Letter	Subshell Name
1	0	*s*	1*s*
2	0 1	*s* *p*	2*s* 2*p*
3	0 1 2	*s* *p* *d*	3*s* 3*p* 3*d*
4	0 1 2 3	*s* *p* *d* *f*	4*s* 4*p* 4*d* 4*f*

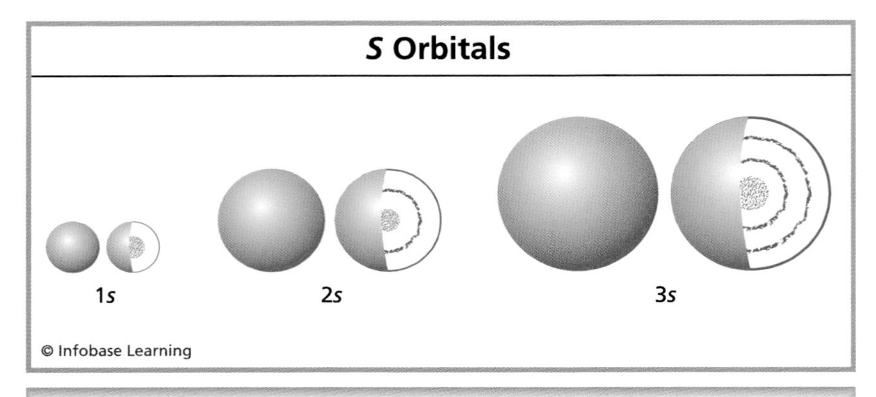

Figure 6.1 This is an image of the spherical shapes of the first three *s* orbitals. The shaded surfaces of the spheres indicate locations where there is a high probability of finding an electron.

Orbital shapes represent electron probabilities. The *p* orbitals are dumbbell shaped, and all but one *d* orbital have four lobes (Figure 6.2). The chance of finding an electron within the boundary of an

Table 6.4 Allowable Quantum Numbers for the First Four Energy Shells

Principal Quantum Number (*n*)	Angular Momentum Quantum Number (ℓ)	Subshell Designation	Magnetic Quantum Number (mℓ)	Number of Orbitals
1	0	1*s*	0	1
2	0 1	2*s* 2*p*	0 -1, 0, +1	1 3
3	0 1 2	3*s* 3*p* 3*d*	0 -1, 0, +1 -2, -1, 0, +1, +2	1 3 5
4	0 1 2 3	4*s* 4*p* 4*d* 4*f*	0 -1, 0, +1 -2, -1, 0, +1, +2 -3, -2, -1, 0, +1, +2, +3	1 3 5 7
Limits of Quantum Numbers				
n = 1,2,3...	ℓ = 0,1,... (n-1)		m_ℓ = -ℓ..., 0, ... +ℓ	

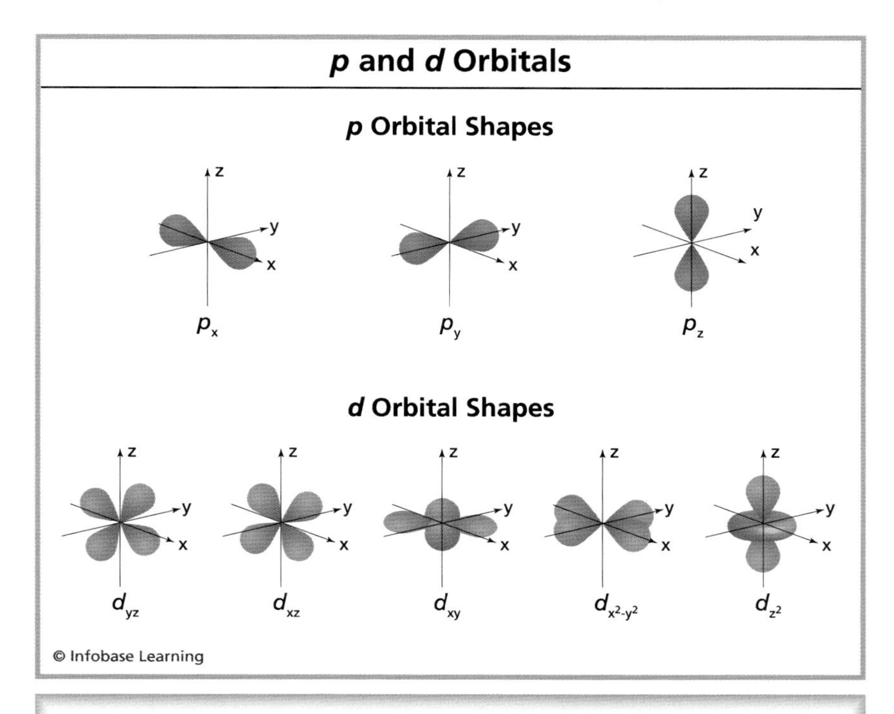

Figure 6.2 The *p* orbitals have two lobes, while most of the *d* orbitals have four lobes.

orbital is approximately 90%. The shaded areas are regions where an electron is most likely to be found.

The last quantum number was proposed to account for the splitting of spectral lines, the final structure that had puzzled Bohr and his colleagues. Several physicists had a hand in trying to solve this problem. However, consensus was not reached until 1925 when Wolfgang Pauli suggested that a new quantum property and number were needed to explain the phenomenon. At the time, the electron was considered to be a particle, and scientists called this new property "spin," usually designated as m_S. The **spin quantum number** has only two possible values: +1/2 or -1/2. It is usually depicted as an arrow pointing up or down.

The spin quantum number brings up a question. What physical features of the atom do the quantum numbers represent? The answer is ambiguous because of the history associated with the quantum numbers.

Quantum numbers were first developed for the Bohr atom when electrons were thought to be charged particles orbiting a nucleus. As mentioned earlier, the principal energy quantum number corresponds to the average energy of the electrons in a shell of the Bohr atom. The angular momentum quantum number is associated, not too surprisingly, with the angular momentum of an electron in an elliptical orbit. The magnetic quantum number is related to the behavior of electrons in a magnetic field. And spin could be visualized as an electron spinning on its axis.

Because wave theory gives more accurate description of the subatomic world, it replaced Bohr's model, and the meaning of quantum numbers became less certain. Can a wave really spin on its own axis? The answer is no. It is sometimes useful to think of quantum numbers as conferring concrete, physical characteristics to an electron. However, quantum properties are only fuzzily related to things in the normal, human-sized world. Thus, electron spin has no ordinary physical meaning. Electrons do not spin like tops—or anything else. The Nobel Prize-winning physicist Richard Feynman hit the nail on the head when he said: "Things on a very small scale behave like nothing you have any direct experience about." And everything scientists have learned about atoms bears out this observation.

BUILDING ATOMS

The principal quantum number establishes the average energy of the electrons in an energy shell. However, electrons in a subshell of a principal energy shell also have different energies. For the n = 3 energy shell, for instance, the energy of the electrons in the 3*s*, 3*p*, and 3*d* subshells are slightly different. To build atoms, it is necessary to know why this is the case.

A helium atom has two protons in its nucleus and two electrons outside of it, twice as many of each as are found in hydrogen. Since positive charges attract negative charges, the nucleus of helium should exert twice as much force on its electrons as hydrogen does. This means it should be twice as hard to remove an electron from a helium atom as it is to remove one from hydrogen. However, this is not the case. Instead of twice as much energy, it takes only about 1.9 times as much.

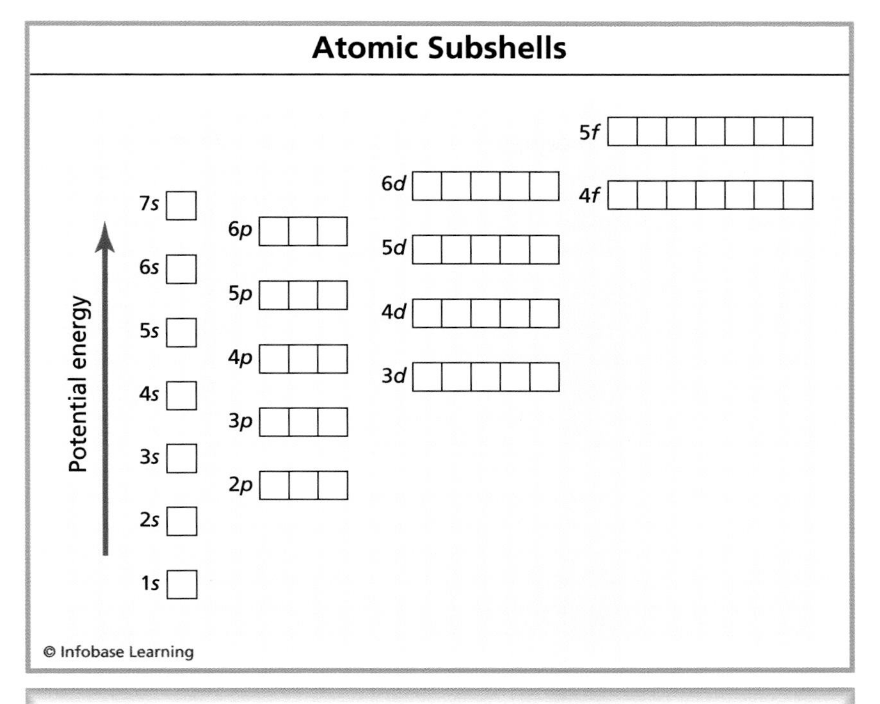

Figure 6.3 Relative energy levels of the atomic subshells are modeled in this image.

Extracting an electron from helium takes less energy than expected because of electron-electron repulsion. The helium nucleus actually does pull twice as a hard on its electrons as a hydrogen nucleus does, but the two electrons in helium also repel one another. The net effect makes an electron in a multielectron atom easier to remove than one would expect if the other electrons were not present.

Figure 6.3 is a diagram of the energy levels of atomic subshells. In some cases, the energy of a subshell in a lower principal energy level is greater than that of a subshell in a higher principal energy level. A 4*d* orbital, for instance, has higher energy than a 5*s* orbital. This is unexpected. It happens because electrons in a 4*d* orbital are disproportionately repelled by the electrons in the inner *s* orbitals. Consequently, it takes less energy to remove an electron from a 4*d* orbital than it takes to remove one from a 5*s* orbital.

Knowing the energy levels of the orbitals enables us to begin building atoms. The lightest atom is hydrogen, which has 1 proton and 1 electron. So, into which orbital should that electron go? The answer, as we have seen, is in a 1*s* orbital. Still, why not a 2*p* or 5*d*? The answer comes from a rule formulated by Niels Bohr back in the 1920s when he was building atoms using his new, quantized atomic structure. This rule is called the **Aufbau principle**. This principle is the first of three rules that are needed to build atoms. It states simply that lower-energy orbitals fill first. Figure 6.3 clearly shows that the 1*s* orbital has the lowest energy. Therefore, the electron must go there. Helium, the next lightest element, has two electrons. According to the Aufbau principle, it, too, would go in the 1*s* orbital.

Electron Configurations Simplified

The modern picture of an atom includes electrons acting as wave and particle. Furthermore, electrons are so elusive, they require probability functions to indicate where they might be. In this picture, an electron's energy and likely position are determined by a set of four abstract quantum numbers that arise from the mathematics of the wave equation, not from physical reality. For these and other reasons, electron configurations are hard to keep straight.

Consequently, for those new to quantum theory, it is sometimes useful to return to the language Niels Bohr used to describe an atom in the early days of its study. What follows is a summary of what shells, subshells, and orbitals are and how they fit together, described in the more concrete language that Bohr himself might have used:

- Electrons in atoms are found in orbitals. No more than two electrons can occupy an orbital and they must have different spin states.
- Orbitals are arranged in energy shells, which are numbered 1, 2, 3 . . . *n*, where *n* is a positive

The next element is lithium, which has three electrons. However, the third electron does not go in the 1*s* orbit. The reason it does not arises from one the most important rules in quantum mechanics. This rule was formulated in 1925 by Wolfgang Pauli, who was awarded a Nobel Prize for it 20 years later. The rule is called the **Pauli exclusion principle**, and it is what makes quantum numbers so crucial to our understanding of atoms.

The exclusion principle states that no two electrons in an atom can have the same set of quantum numbers. The 1*s* orbital has the following set of allowable numbers: $n = 1$, $\ell = 0$, $m_\ell = 0$, $m_S = +1/2$ or $-1/2$. All of these numbers can have only one value except for spin, which has two possible states. Thus, the exclusion principle restricts the 1*s* orbital to two electrons with opposite spins. A third electron

whole number. Usually, the higher the value of *n*, the higher the energy of the electron.

- Shells where *n* is greater than 1 may contain subshells, labeled *s, p, d,* and *f.* These subshells are related to the angular momentum of an electron in an elliptical orbit.
- The energy of an electron in an orbital in one subshell is slightly different from the energy of one in another subshell, even if both subshells have the same value of *n*. This happens because the geometry of an orbital in, say, a *p* subshell is different from the geometry of an orbital in an *s* subshell. The difference in the shapes of the orbitals produces varying degrees of electron-electron repulsion, which affects the energy of the electrons.
- The orbitals in a given subshell have a different orientation in space but have the same energy as the other orbitals in that subshell (except when the atom is in a magnetic field). Thus, all three of the 2*p* orbitals of nitrogen are oriented differently from one another but the electrons in them all have the same energy.

in the 1*s* orbital would have to have a set of quantum numbers identical to those of one of the electrons already there. This is forbidden by the exclusion principle. So, the third electron needed for lithium must go into the next higher energy shell, which is a 2*s* orbital.

The maximum number of electrons allowed for any atom with a principal quantum number of n is the number of orbitals shown in Table 6.4 multiplied by 2. Multiplying by 2 allows a pair of electrons with opposite spins to occupy each orbital.

The final complication in building atoms comes when one reaches carbon. A carbon atom has 6 electrons. To build this atom, the first 2 electrons go into the 1*s* orbital, the second pair in the 2*s* orbital. The fifth electron must go into a 2*p* orbital. However, into which of the 3 2*p* orbitals should the sixth electron go? In the orbital already occupied by the fifth electron or in one of the unoccupied orbitals?

The answer to that question came from the last rule needed to generate the electron configurations of atoms. It was proposed by German spectroscopist Friedrich Hund. **Hund's rule** can be expressed in several ways. The most precise definition is that an atom in a higher net spin state is more stable than one in a lower spin state. To satisfy this rule, the sixth electron in carbon must have the same spin as the fifth one. And the Pauli exclusion principle bars the sixth one from going in an already-occupied *p* orbital. Thus, it must fill an empty *p* orbital.

Knowing these three rules—the Aufbau principle, the Pauli exclusion principle, and Hund's rule—and the energy levels of the subshells shown in Figure 6.3, one can build the correct electron configurations of most atoms.

These electron configurations tell chemists much of what they need to know about the chemical properties of atoms. And, as Rutherford no doubt would have loved to point out, the discoveries that led to those configurations came from physicists: Bohr, Heisenberg, Schrödinger, Pauli, and, of course, Rutherford himself.

Despite these advances, the quantum theory itself was still in a spot of trouble. It had two competing mathematical approaches. Which one of them, matrices or waves, was correct?

7

Extending the Theory

Although scientists were quick to adopt Erwin Schrödinger's wave equation to solve practical problems, theoreticians were troubled. They now had two systems of quantum mechanics. Werner Heisenberg was pushing his matrix approach, while Schrödinger, not surprisingly, advocated for waves. Both systems gave the same correct answers, but the mathematics that supported them were miles apart. Which system was right? Waves? Matrices? Both?

The two men primarily responsible for developing the systems were not fond of the other's ideas. "I am rock-solid convinced," wrote Heisenberg to a colleague, "of the inconsistencies of . . . the quantum mechanics presented by Schrödinger." Schrödinger, on the other hand, felt that his wave theory was less abstract and easier to understand than Heisenberg's matrix-based approach. "I was discouraged, if not repelled," he wrote, "by what appeared to me as a very difficult nettle of transcendental algebra, defying any visualization."

Nevertheless, Schrödinger worked to understand Heisenberg's matrix mathematics. Within months of publishing his first paper on wave theory, he had reached a conclusion. The two approaches were equivalent, he said, two ways of getting the same result. However, a more rigorous analysis was just around the corner. It came from another prominent figure in the development of quantum theory. His name was Paul Adrien Maurice Dirac.

PAUL A.M. DIRAC

Much of what we know about quantum mechanics today was developed during an eighteen-year period. It began with Bohr's description of the quantized atom in 1913 and lasted until about 1931. It was a feverish era, dominated by scientific giants, all of whom were men, all of whom were physicists, and all of whom were acquainted with one another. Bohr was the heart and soul of the push to understand the strange world of the atom. Einstein was the movement's intellectual leader. Seven others—all Nobel Prize winners but not quite as well known outside of scientific circles as Bohr and Einstein—also had starring roles. The contributions of six of them—Planck, de Broglie, Heisenberg, Born, Pauli, and Schrödinger—have been discussed. The seventh man was Paul Dirac, whom Bohr described as having "the purest soul" of all physicists.

The phrase *purest soul* was not a reference to Dirac's religious propensities. In fact, an acquaintance wrote that he "tended toward atheism." Actually, Bohr was alluding to Dirac's way of working.

Just as scientists come from all walks of life, from bookbinder's apprentice to prince, they also have different ways of attacking a problem. Some of them look for anomalies, or data that do not fit the current theory, and then propose hypotheses to explain the discrepancy. Other scientists design experiments to find new facts, and then use the results to advance science or create new products. Some scientists—Einstein being one—think about how the universe should operate and dream up theories that others can test. Dirac did not fit any of those molds. He had a completely different approach.

Dirac loved mathematics. He believed in playing with equations to see where the math led him. Sometimes, it led to new physics. Research, Dirac wrote, "is simply a search for pretty mathematics. It may turn out later that the work does have an application. Then one has good luck."

Of course, the mathematics Dirac played with are unfamiliar to most people, even to many physicists. He lived in a land of Laplacian operators, quaternions, and Riemannian geometry; of Clifford algebras and four (or more) dimensional manifolds. Little in Dirac's early background prepared him for these rarified mathematics.

He was born in Bristol, England, in 1902, the same year that Albert Einstein started work at the patent office in Bern. His father

Figure 7.1 In 1931, Paul Adrien Maurice Dirac (*above*) predicted the existence of antimatter; he was proven correct the following year by American physicist Carl Anderson.

taught French and insisted that Paul speak to him only in that language. "Since I found that I couldn't express myself in French," Dirac recalled, "it was better for me to stay silent . . ."

Not surprisingly, Dirac grew up to be a man of few words. In fact, his stinginess with them was legendary. He once argued that the phrase "not to be published in any form" was wordy because the words "in any form" were redundant. Not surprisingly, the few words Dirac did utter (or write) were always well thought out. "One never should begin a sentence," he said, "until one knows what the end of it will be."

Pushed by his father, who was concerned about his son's prospects for a job, Paul studied electrical engineering at the University of Bristol. His father's judgment about the job market in Britain proved to be less than accurate, and Paul was unable to find a decent job after he graduated. For two years, he lived at home and studied mathematics at Bristol. Finally, he got what he wanted—an unusually generous scholarship that enabled him to enter Cambridge to study mathematics and physics. Nine years later, he was appointed to the Lucasian Chair of Mathematics at Cambridge, the prestigious post once held by Isaac Newton (and, more recently, by Stephen Hawking). One year later, in 1933, he shared the Nobel Prize for Physics with Erwin Schrödinger.

What did Dirac do to justify such honors? Among other things, he bridged the gap between wave mechanics and matrix methods. He attacked the problem by defining the noncommutative variables as operations rather than numbers. Thus, the baffling result of multiplying A times B and getting a different answer than multiplying B times A, is no longer surprising. Operations usually are noncommutative. Putting on your raincoat first and then your shirt provides an entirely different result than putting on the shirt first and then the raincoat.

Following one particular mathematical spoor led Dirac to a more general expression of quantum theory. The new approach demonstrated the equivalence of Heisenberg's matrix mechanics and Schrödinger's wave equations in far more rigorous way than Schrödinger's earlier work had. Dirac's more abstract formulation was, Einstein said, "the most logically perfect presentation of quantum mechanics."

Dirac went on to make many more contributions to quantum theory. One of these deserves a special mention because of a famous prediction that came out of it. In 1928, Dirac added relativistic effects to the wave equation of an electron. The result, now known as Dirac's equation, predicted the existence of a new particle, which Dirac called an antielectron. The new particle should have the same mass as an electron but carry a positive charge instead of a negative one.

This prediction was so outrageous that Dirac's colleagues had trouble accepting it. Even Dirac himself was ambivalent and searched for other explanations. No one had ever detected such a particle or even dreamed of its existence. However, finally, in 1931, Dirac unequivocally predicted the existence of the antielectron. Within a year, the American physicist Carl Anderson discovered the particle, which he would later name the **positron**. Dirac's prediction of an elementary particle before there was any experimental evidence of its existence is one of the great triumphs of quantum theory.

Surprisingly, this exotic new particle turned out to be useful. Positron emission tomography (PET) is used in hospitals every day to detect, monitor, and guide the treatment of many diseases, from cancer to Alzheimer's.

THE UNCERTAINTY PRINCIPLE

After Dirac proved the equivalence of the matrix and wave approaches to quantum mechanics, Bohr and Heisenberg began to ponder the central conundrum posed by the great discoveries of the previous few years—the wave-particle duality of light. The double-slit experiment showed clearly that light was a wave. The photoelectric effect showed just as clearly that it was a particle. Einstein had accepted this duality. Now, beginning in September 1926, Bohr and Heisenberg were trying to figure out this wave-particle duality.

Heisenberg recognized that one could visualize a wave. One could also visualize a particle. However, something that could be either wave or particle made no sense in the ordinary world. Nevertheless, wave-particle duality was clearly part of the subatomic world. This realization took Heisenberg back to his original mathematical approach to the quantum, which led to a new and totally unexpected principle of quantum theory.

Even the mathematically minded Heisenberg found matrix algebra strange and difficult to master. Still, the accuracy of its predictions had been confirmed in numerous experiments. Was there something odd in the matrix method that might shed light on wave-particle duality? One of the strangest aspects of matrix multiplication had been the noncommutativity of matrices. Both the momentum, p, and position, q, of an electron could be written as matrices. Therefore, according to rules of matrix algebra,

$$pq \neq qp$$

or

$$pq - qp \neq 0$$

This result makes absolutely no sense in classical physics. Multiplying the momentum of an object by its position always gives the same result as multiplying position by momentum. That is

$$pq - qp = 0$$

Yet according to Heisenberg's quantum theory, the two products were not equal. How could that be? Both momentum and position are measurable, experimental quantities. To Heisenberg, the inequality of pq and qp suggested that the problem might be one of measurement. Just how accurately could p and q be measured? Could the inequality be due to uncertainty in measuring the two quantities?

Heisenberg answered those questions in 1927 in one of the most important scientific papers published in the twentieth century. He begins with the statement that certain quantities "can be determined simultaneously only with a characteristic indeterminacy." Take, for instance, the momentum and position of an electron. Heisenberg stated the **uncertainty principle** for these two variables as follows: "The more precisely the position is determined, the less precisely the momentum is known in this instant, and vice versa."

Heisenberg attributed this uncertainty to problems with observation. To measure the position or speed of an electron, one must shine light (or electromagnetic radiation of some other frequency) at the particle. As Einstein showed in his analysis of the photoelectric effect, electromagnetic radiation has momentum associated with it, which is why it can knock an electron out of a metal plate. Thus, any measurement must disturb the thing you are trying to measure. The disturbance is trivially small if the thing is a large macroscopic body such as a person or a car. However, if the object is small enough, as an electron is, then the collision between the measuring photon and the electron will affect the electron's momentum and position.

Further, to measure a particle's location more precisely, one must probe with electromagnetic radiation of shorter and shorter wavelengths, which means higher and higher frequencies. However, Planck's law says that higher frequency radiation is more energetic. For this reason, higher resolution measurements will, on average, impart more momentum to the object being measured. This increases the uncertainty of the momentum measurement. And if you try to make a measurement with longer wavelength, or lower energy radiation, the momentum imparted to the particle being measured will decline. However, the uncertainty about its position will increase.

The uncertainty principle can be expressed mathematically in several ways. One of the more common is

$$\Delta x \Delta p \geq h/4\pi$$

where Δx is the uncertainty in the x coordinate of a particle, Δp is the uncertainty in the x component of its momentum, and h is Planck's constant. The symbol $\geq$ means greater than or equal to.

Max Born's statistical interpretation of the wave function led to the conclusion that one cannot pinpoint the location of an electron in an atom. One can only know the probability of where it might be. Now, Heisenberg pointed out that even if we could know exactly the location of an electron, we could not know much about its momentum or velocity. Meanwhile, back in Copenhagen, Niels Bohr was preparing to answer the question about wave-particle duality, the question that started Heisenberg thinking about uncertainty. The answer would add even more disturbing ideas to the quantum theory.

THE COPENHAGEN INTERPRETATION

The heart of the quantum theory was developed in the 1920s during a frantic period of discovery and hypothesis. By 1927, Niels Bohr had published little but, nevertheless, was asking hard questions. How does quantum mechanics fit in with classical physics, which is the physics of the steam engines, electricity, and bridges on which society depends? What does it mean when we say an electron can be wave or particle? What can we know about the properties of

(continues on page 88)

Important Events in Quantum Theory

Between 1913 and 1931, advances in the quantum theory came so fast and furious that it is hard to keep up with what happened when. The summary of important dates and events listed below should help:

1913 Bohr publishes a paper on the quantum atom.

1921 Institute for Theoretical Physics acquires first building.

1921 Pauli becomes an assistant to Max Born at Göttingen. He is replaced by Heisenberg in 1924.

1923 De Broglie conceives his idea about the wave nature of matter for his Ph.D. thesis.

1923 Heisenberg gets his Ph.D. working for Max Born.

1923 Arthur Compton shoots X-rays at electrons, proving that electromagnetic radiation is composed of photons.

1924–1926 Heisenberg works with Bohr.

1924–1925 Pauli suggests that electrons must have a new quantum property, later called "spin."

1924 In November, De Broglie defends his wave-particle thesis.

1925 Early in the year, De Broglie publishes wave-particle hypothesis.

1925 In May, Heisenberg works out his matrix theory of quantum mechanics and, in July, gives it to Max Born in Göttingen to submit for publication.

1925 Pauli proposes the exclusion principle.

1925 In November, Born, Heisenberg, and Jordan submit comprehensive paper on matrix mechanics.

1925 On December 1, Dirac publishes paper on quantum algebra that includes matrix mechanics.

1926 On January 27, Schrödinger's first paper on wave mechanics is published.

1926 On February 23, Schrödinger's second paper on wave mechanics appears.

1926 In March, Schrödinger establishes that Heisenberg's approach to quantum theory is mathematically equivalent to his own.

1926 In the spring, Max Born shows how the wave equation gives the probable location of an electron.

1926 On May 10, Schrödinger's third paper on wave mechanics is published.

1926 G.N. Lewis introduces the word *photon* to describe a particle of light.

1926 During the fall, Bohr and Heisenberg meet daily.

1926 In September, Schrödinger visits Bohr and Heisenberg participates in the discussions. During these meetings, he begins to explore whether uncertainty is built into the equations of quantum mechanics.

1926–1927 Bohr and Heisenberg collaborate. Heisenberg proposes the uncertainty principle.

1927 On January 1, Dirac publishes transformation theory, which shows how the several formulations of quantum mechanics are equivalent.

1927 In the spring, Heisenberg publishes the uncertainty principle.

1927 Clinton Davisson and Lester Germer confirm de Broglie's hypothesis about the wave nature of the electron.

1927 In September, Bohr presents the Copenhagen interpretation of the quantum theory at a conference in Italy.

1927 In October, Bohr presents the Copenhagen interpretation of quantum theory at the Solvay Conference in Brussels.

1928 During February and March, Dirac adds relativity to the wave formalism of the electron and predicts the existence of an antiparticle.

1932 Anderson produces the first evidence of the positron.

(continued from page 85)
subatomic particles in light of Max Born's probabilities and Heisenberg's uncertainties?

The Copenhagen interpretation was an attempt to answer those questions. It was stitched together by Bohr with help from Heisenberg, Pauli, and others. The idea was to describe quantum processes in everyday language, the language used to explain the behavior of normal-sized objects. The result is as much philosophy as science. Even today, physicists disagree on exactly what the Copenhagen interpretation is and what it means.

The first leg supporting the amorphous platform that became known as the Copenhagen interpretation was suggested by Bohr before the quantum theory was fully developed. He was looking for a way to reconcile the discontinuities found in the behavior of small objects with the smoothly continuous behavior of large ones. He wanted to understand the relationship between quantum rules and the laws of classical physics. To ensure agreement between the two realms, Bohr came up with the **correspondence principle**.

In describing the behavior of large systems, both classical physics and quantum mechanics should give approximately the same results. In other words, there is no need to discard Newton's laws because of the quantum theory. With large, human-sized systems, both sets of laws must give the same results. The correspondence principle has proved useful in aspects of science other than the quantum theory. It can be invoked, for example, to explain why the equation describing the relativistic energy of a body moving at a velocity well below the speed of light must (and does) reduce to give the classical physics equation for kinetic energy.

The correspondence principle solved a big conundrum posed by the quantum theory. How can two entirely different sets of laws—those from classical physics and those derived from quantum theory—give the same results when applied to ordinary objects? The answer, according to Bohr, was that all quantum rules must reduce to the laws of classical physics when applied to large systems. Those rules that do not are invalid.

All his life, Niels Bohr struggled with the written word. Writing papers was agonizing for him and involved many labored revisions. However, he was creative with language. The use of the word *correspondence*, which means "agreement," in his correspondence principle was a perfect fit. Similarly, when Bohr needed a word to describe wave-particle duality, he came up with another good one: *complementarity*.

To Bohr, the word meant that the two seemingly mutually exclusive descriptions of quantum entities—light, for instance—as wave or particle complement (or complete) one another. Both descriptions are needed to fully characterize light.

Remember the two key experiments with light? The double-slit experiment showed that light was a wave. The photoelectric effect proved equally conclusively that light was a particle. Well, one wonders, could light exhibit both wave and particle characteristics at the same time? Surprisingly, some poets and writers think so; for example, the distinguished American author Ellen Gilchrist titled one of her books *Light Can Be Both Wave and Particle*.

Unfortunately, Ms. Gilchrist had it wrong. No experiment has ever showed light acting as both wave and particle. In fact, light cannot be simultaneously both wave and particle. It is either wave *or* particle. That distinction raises an important question: What determines the behavior of light (or subatomic particles such as electrons)? What makes it behave like a particle on some occasions and like a wave on others?

The incredible answer is that the behavior of light depends on the experiment that is performed. This remarkable conclusion, reached by Bohr as part of the Copenhagen interpretation of the quantum theory, presented (and still presents) a huge problem for many physicists and philosophers. Does nature actually depend on what experiment one chooses to run? Is there no objective reality, a reality independent of the experimenter? Actually, according to Bohr and his disciples, no.

The Copenhagen interpretation, as articulated by Bohr and Heisenberg, was (and is) a startling new take on the nature of the physical world. Heisenberg's uncertainty principle and Max Born's hypothesis that the wave nature of electrons made it impossible to precisely locate them (or any other quantum entity) eliminated the deterministic, predictable world of classical physics. The new world, the quantum world, was ruled by uncertainty and probabilities.

Bohr presented his version of the Copenhagen interpretation (of which there were and are many variants) at a scientific conference at Lake Como, Italy. Many of the world's leading scientists were

Figure 7.2 The 1927 Solvay Conference brought together many of the greatest minds in science to discuss the newly formulated quantum theory. Seventeen of the 29 attendees were or later became Nobel Prize winners, including Albert Einstein (*bottom row, center*), Edwin Schrödinger (*top row, sixth from the left*), Paul A.M. Dirac (*middle row, fifth from left*), Niels Bohr (*middle row, first on the right*), and Marie Curie (*bottom row, third from left*).

present but not Einstein. Bohr's talk provoked discussion but little controversy. That would come the following month when he presented his findings at one of the most famous scientific conferences ever held: the Solvay Conference that convened in October 1927 in Brussels, Belgium.

The topic was "Electrons and Photons," and everyone who was anyone in the world of physics was there. Max Planck, Max Born, and Marie Curie attended, as did Paul Dirac, Erwin Schrödinger, and Wolfgang Pauli. Louis de Broglie, Arthur Compton, and many other distinguished physicists also came. Nevertheless, the stars of the show were Bohr and Einstein.

Einstein could not accept the conclusions that Bohr presented, which laid out the case for the Copenhagen interpretation. It was the beginning of a long, hard-fought, but friendly, argument between the two titans of science that would last until Einstein's death in 1955.

Einstein disagreed with Bohr's idea that the universe operated on chance and probabilities. "In any case," Einstein wrote to Max Born, "I am convinced that He [God] does not throw dice." Upon hearing this comment, Bohr is supposed to have replied, "Einstein, do not tell God what to do with his dice."

Some of the concepts associated with the Copenhagen interpretation proved to be a tough pill for many other scientists to swallow, too. Some of them were influenced by Einstein. "I attach only a transitory importance to this interpretation," he said. "I still believe

What Are Solvay Conferences?

The Solvay Conferences (or, in French, *Conseils Solvay*) were the brainchild of Ernest Solvay, a man of great wealth and a passion for science. In 1911, he organized a meeting of the best-known physicists and chemists of the day. The goal was to "promote research, the purpose of which is to enlarge and deepen the understanding of natural phenomena."

Solvay conferences continue to this day. Participation is by invitation only and is limited to the most accomplished researchers. The conferences are held on a three-year cycle, with physics being the focus of one conference. No conference is held the following year, and the year after that focuses on chemistry. Then, the cycle starts again. So far, there have been 20 conferences on chemistry and 23 on physics. However, although they continue to attract the top talent, it is unlikely that any of today's conferences have featured a cast that could outshine the star-studded group that assembled for the 1927 meeting.

in the possibility of a model of reality." Today, however, most physicists side with Bohr and the Copenhagen interpretation.

It is no wonder that even now, many people have trouble subscribing wholeheartedly to the quantum theory—it is strange and hard to accept. As Bohr himself once wrote, "If quantum mechanics hasn't profoundly shocked you, you haven't understood it yet." Nevertheless, experiment after experiment has proved that the quantum theory accurately predicts the behavior of electromagnetic radiation, electrons, atoms, and more.

Accounting for the behavior of tiny particles may seem too abstract to affect humans who live on a vastly larger scale. Nevertheless, big objects are made up of small ones, and small ones of even smaller ones. Thus, quantum effects at the atomic and subatomic levels can have dramatic consequences in the larger world and on the way we live. The next chapter will show how the hard-won knowledge generated by the scientists who forged the quantum theory has changed society.

8

Why Does Quantum Theory Matter?

Although work on the quantum theory continues to this day, the fundamentals were put in place during a 30-year period that began with Max Planck's equation in 1900. The theory's history is dominated by scientific giants who produced startling insights into the world of the very small—but so what? Sure, it explained the photoelectric effect and blackbody radiation, but does it give an accurate picture of the ordinary world? In other words, is the quantum theory really right? And if it is, how has it affected our lives? Finally, why should anyone bother to learn about it, anyway? This chapter will explore those questions.

First, is the quantum theory right? Scientific hypotheses are usually labeled as theories or laws, such as "quantum theory" and "Newton's laws." It is tempting to conclude that laws are better established and more likely to be correct than theories. However, nothing could be further from the truth. In fact, under certain conditions, Newton's laws are less accurate than those of the quantum theory. So, how well do the predictions of the theory match the experimental results?

Nobel Prize-winning physicist Richard Feynman weighed in on this question in his book *QED: The Strange Theory of Light and Matter*. He compared the experimentally measured magnetic moment of an electron to the value that is calculated using quantum theory.

The numbers were so close, Feynman wrote, that "If you were to measure the distance from Los Angeles to New York to this accuracy, it would be exact to the thickness of a human hair."

Two physicists, Bruce Rosenblum and Fred Kuttner, assess the accuracy of the theory in their book, *Quantum Enigma*. "Quantum theory," they state boldly, "is the most stunningly successful theory in all of science. Not a single one of its predictions has ever been wrong."

However, like any other scientific theory or law or hypothesis, one cannot say for sure that the quantum theory is correct. Data could come in tomorrow to disprove it or any other theory. So far, though, it is as sound as a scientific theory can be. Based on everything that scientists know today, the quantum theory provides an accurate and useful view of the universe. It accounts for a wide range of phenomena, such as atomic spectra, for which Newtonian physics could offer no convincing explanation. The theory has also fathered practical applications that have transformed modern life.

Some of these are so obvious and so thoroughly examined, it hardly seems worthwhile to explore them further. Nuclear power, for example, depends on our knowledge of the quantum. As Richard Rhodes points out in his exhaustive history of the atomic bomb, "Atomic energy requires an atom." And atoms made no sense until Niels Bohr used quantum theory to show why they behave as they do. Today, two products of the quantum revolution have woven themselves so deeply into the fabric of modern life that they deserve examining.

STAR WARS, STIMULATED EMISSIONS, AND ALBERT EINSTEIN

In 1983, U.S. president Ronald Reagan proposed the Strategic Defense Initiative. Nicknamed "Star Wars," its core idea was to produce a beam of electromagnetic radiation so powerful that it could shoot down incoming enemy missiles. The stream of electromagnetic radiation was to be generated by a **laser**. What does such a mighty weapon have to do with the quantum theory, a theory of the very small? Well, as it turns out, almost everything.

The idea for a laser was first suggested by Albert Einstein in 1917 in a scientific journal. His paper elaborated on a concept that Einstein had mentioned in a letter a few months earlier. "A splendid light has dawned on me," he wrote, "about the absorption and emission of radiation." In his paper, Einstein showed that atoms can interact with photons in three ways. Two of the ways—the ordinary absorption and emission processes (Figures 8.1A and B)—are well known. The illustrations show a hypothetical atom with two possible energy states. The atom can move from the lower energy state (its ground state) to the higher one (its excited state) by absorbing a photon of energy. It decays from the excited state to the ground state by emitting a photon. Both the emitted and absorbed photons have exactly the same energy (or frequency, as determined by Planck's equation, $\Delta E = hf$). However, Einstein's "splendid light" revealed a new type of emission that is called a **stimulated emission**.

In stimulated emissions, a photon interacts with an atom already in an excited energy state. Einstein showed that if the incident photon has the same energy as the difference between the atom's two energy states, ΔE, then the excited atom will emit a photon and drop to a lower energy level. The result of this interaction is an atom in the ground state plus two photons with the same wavelength, as shown in our hypothetical atom in Figure 8.1C.

It is tempting to think of stimulated emissions as a perpetual motion machine, a system that refutes the **conservation of energy** law. After all, you put in one photon and you get two out. Unfortunately, like all perpetual motion schemes, the analysis in incomplete. Energy must be added to the system to excite the atoms to the higher-energy state. So, more energy is required for stimulated emissions than comes out.

While stimulated emissions systems are not perpetual motion machines, they do act as amplifiers. One photon produces another photon and yet another, amplifying the number of photons. The word *laser* comes from a literal description of the device. It is an acronym for Light Amplification by Stimulated Emission of Radiation.

Although lasers dominate stimulated-emission devices today, the first workable gadget based on Einstein's idea was not a laser but a maser. Here, the m stands for microwave because masers amplify microwave radiation rather than light.

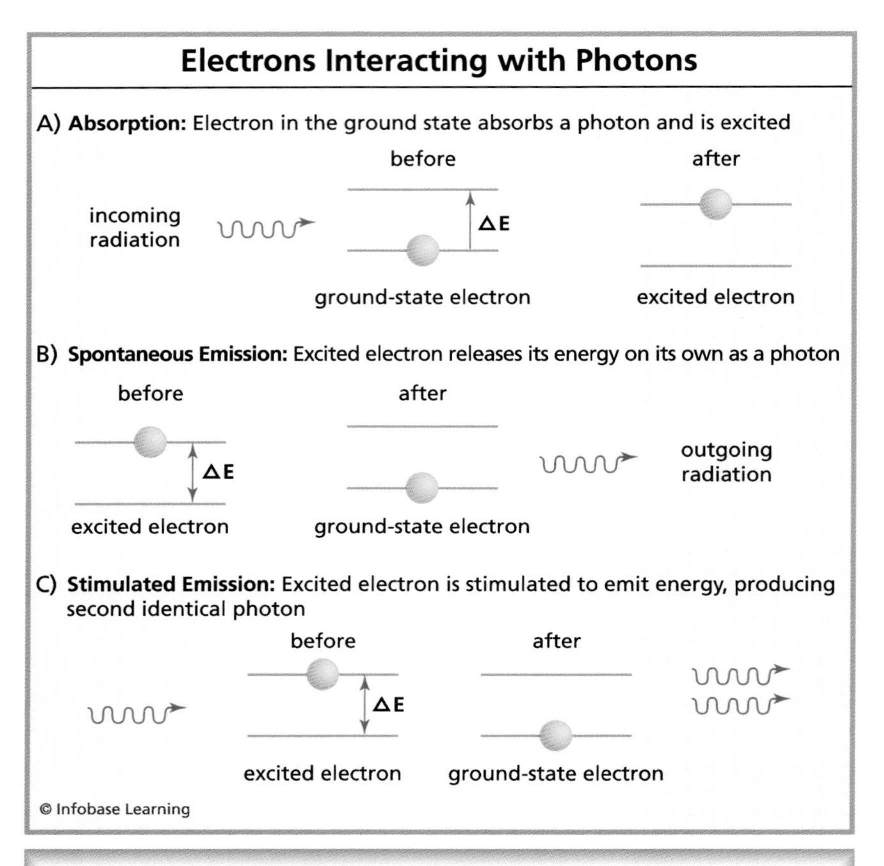

Figure 8.1 An electron in the lower energy state can absorb a photon and jump to the higher level (A), an electron in the higher energy state can spontaneously emit a proton (B), or an electron in the higher energy state can be stimulated to emit a photon (C).

MASERS AND LASERS

It took almost 40 years to turn Einstein's startling insight into a workable maser. In the spring of 1954, Charles Townes, a physics professor at Columbia University, overcame numerous sticky engineering problems to build (with help from his graduate students) the first working maser. "The output was a mere 10 billionth of a watt," according to science writer Jeff Hecht, "but it was enough to show it worked." It was also enough to win Townes the Nobel Prize for physics in 1964. He shared the award with Aleksandr Prokhorov and Nikolai Basov, who had published important papers on the theory of masers.

In 1958, Townes and his colleague Arthur Schawlow followed up Townes's work with the maser in a new paper, "Infrared and Optical Masers." The paper gave the theoretical outline for constructing a laser, a higher-energy version of a maser that emitted electromagnetic radiation in visible range. After the paper was published, the race to build the first laser was on. Fame and fortune would certainly be heaped upon the winner.

As it turned out, the first person to build a laser was not an established academic surrounded by graduate students and technicians, but an engineer turned physicist named Theodore Maiman, who was working for the Hughes Aircraft Company. In 1960, Maiman published a paper describing the first working laser. Unlike Townes, he did not get a Nobel Prize for his breakthrough, but he was inducted into the National Inventors Hall of Fame. (The company he formed to produce lasers was acquired by chemical giant Union Carbide.)

Maiman's first laser used a synthetic ruby to produce a beam of monochromatic, coherent, highly directional red light. The device consisted of a flash lamp and a ruby crystal shaped like a tube with silvered circular disks at both ends. The flash excites some of the chromium ions, which are present in small amounts in the ruby crystal. When these ions return to the ground state, they emit red light. This light stimulates other excited chromium ions to produce even more photons. The light bounces back and forth between the silvered disks, generating more and more photons with each trip.

Although both of these disks can act as mirrors, one of them is only partially silvered. What finally emerges from the partially silvered mirror is a pulse of intense laser light. Although it takes a while to read this passage describing the process, the event itself is lightning fast. The photons in the laser are traveling at the speed of light, and only about one-thousandth of a second elapses from lamp flash to laser pulse.

The collimated feature of laser light enables scientists to do tricks that would be impossible with ordinary light. A beam of light from a lightbulb, even a very powerful bulb, will spread out. The light from a lighthouse can be seen for maybe 20 miles (32 km). At distances greater than that, the beam is so dispersed that it is impossible to see.

A laser beam, on the other hand, can be bounced off of a reflector on the Moon and still be visible. In fact, by measuring the roundtrip time for a laser beam to travel from Earth to Moon and back again, one can calculate the distance to the Moon that is accurate to less than 1.2 inches (3 centimeters). That is pretty tight measuring, considering that the average distance from Earth to the Moon is about 238,857 miles (384,403 km).

The Star Wars project did not succeed, but lasers are employed these days for many other uses. Diode lasers are used for pointers; gas lasers are used in supermarkets to scan your purchases and relay the information to a computer that will tell you how much money you have spent; and carbon dioxide lasers are indispensable for certain types of welding. Other lasers are used in surveying, in fiber optic cables, and for eye surgery.

Laser Light: Monochromatic, Coherent, and Collimated

The light—or electromagnetic radiation of any frequency—produced by a laser is quite different from ordinary light, such as that from the Sun. As Isaac Newton demonstrated centuries earlier, ordinary white light is a mixture of colors. Each color arises from radiation of a different wavelength. This means ordinary light is composed of radiation of varying wavelengths. Because laser light is a result of stimulated emissions, it is monochromatic. Every photon has the same wavelength.

In addition to being monochromatic, laser light is coherent. That is, every wave is in step with every other wave, so their peaks and troughs exactly coincide. Furthermore, because the beam bounces back and forth in a nearly perfect cylindrical cavity, such as a ruby crystal, laser light is collimated, meaning that it is very tightly focused. The result is a pointer beam that makes a spot smaller than a dime on a projection screen in a lecture room, and a column of light that can reach the reflectors that were set up on the Moon by the Apollo astronauts.

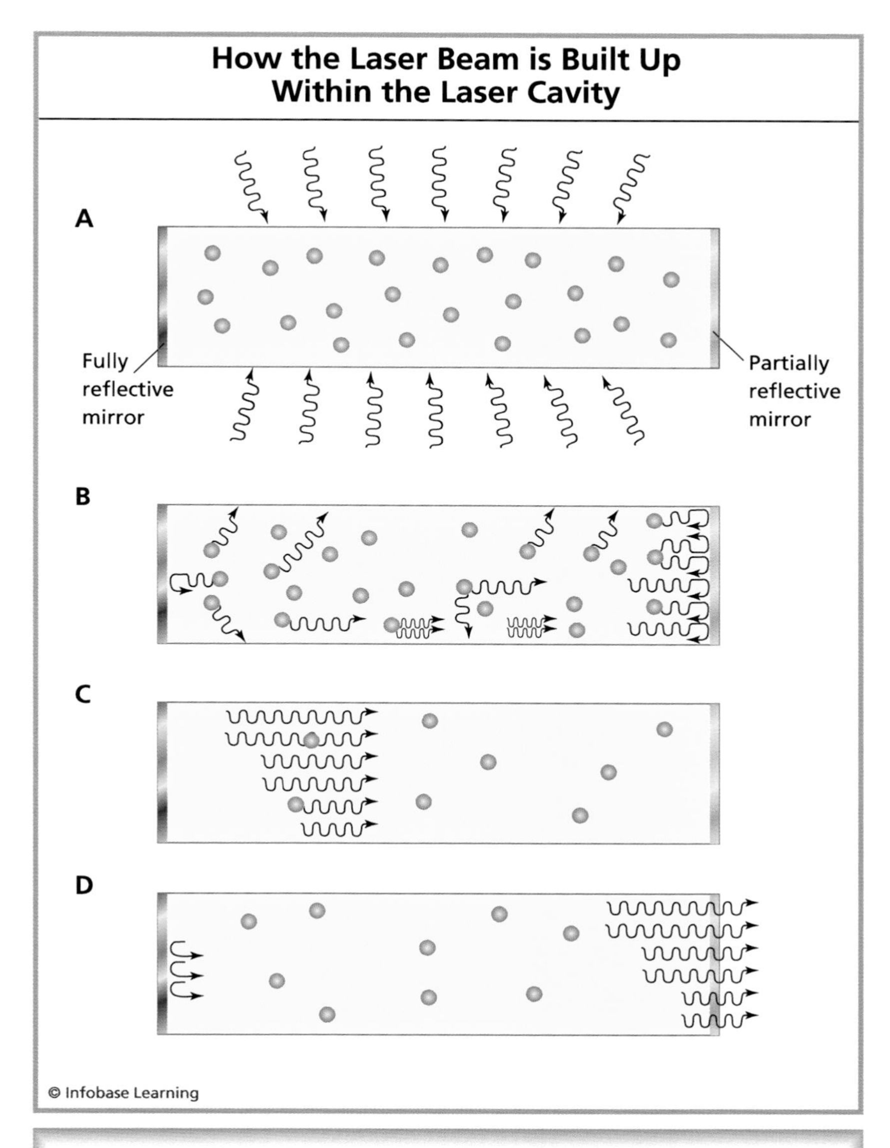

Figure 8.2 When a laser beam is built up in a cavity (A), the laser medium absorbs energy and begins emitting photons (B). After stimulated emission, the photons form a coherent wave (C), and exit upon reaching high intensity (D).

None of these useful devices would have been possible without the quantum theory, but applications of the theory extend well beyond lasers. In fact, they pervade modern life. And nowhere is that pervasiveness more apparent than in our last example of the practical application of quantum theory—the ubiquitous **transistor**.

TOWARD THE TRANSISTOR

Transistors are made from **semiconductors**. To understand semiconductors, one must first know something about electrical conductors. Conductors are substances—usually metals—that readily pass electricity. The electrons in a metal are free to move compared to those in an **insulator** or a semiconductor. When you attach a copper wire to a battery and complete a circuit, current will flow. When you attach a nonmetal, such as a string, nothing happens. Why are metals good conductors of electricity? The answer takes us back to quantum theory.

A central tenet of quantum theory holds that free atoms have discrete energy states. However, when atoms are packed into a solid, the individual orbitals of the billions and billions of atoms overlap, creating huge numbers of orbitals with energy levels that are very close together. These groups of orbitals are called energy bands. However, electrons in those bands must obey the Pauli exclusion principle. So, each orbital can have only two electrons in it. However, the orbitals are so close in energy that electrons can move easily from one orbital to another.

Solids can have more than one energy band. The band with the lowest-energy orbitals is called the valence band. It is almost filled with an atom's outermost or valence electrons. The electrons in the valence band are bound to atoms and are unavailable to conduct electricity.

Another band, called the conducting band, is composed of orbitals with energy states greater than those in the valence band. Dividing the two bands is a line called the **Fermi energy** (or **Fermi level**), named after Enrico Fermi, the distinguished Italian physicist. Using quantum theory, Fermi showed that no electrons can exist above the Fermi level at **absolute zero**. At room temperature, though, thermal energy boosts a few electrons in some solids to energy states above the Fermi level. The Fermi level plays a critical role in the electrical properties of solids because only those electrons above or very near the Fermi level are free to conduct a current.

The position of the Fermi level and the amount of separation between the valence and conducting bands determines the conductivity of a substance. In metals, the valence and conducting bands overlap, and the Fermi level lies in the overlap region. Thus, conductors have many electrons in the conduction band.

These electrons give copper and the other metals their high electrical conductivity. When you push an electron in one end of a metal wire, another electron from an almost identical orbital will pop out at the other end. The orbitals are so close in energy that only a small push is required for this process, making metals highly conductive and the preferred material for power lines.

Insulators, on the other hand, have a gap between the valence and conduction bands, and the Fermi level lies between them. Thus, insulators have almost no electrons available to conduct electricity. This is why America's electrical grid is connected by metal wires rather than strings.

Semiconductors, as the name implies, fall somewhere between conductors and insulators. The valence and conduction bands do not overlap as they do in conductors. However, in semiconductors, the two bands are much closer in energy than they are in insulators, as shown in Figure 8.3. The closeness of the bands makes it possible for an electron to jump from one to another. This gives semiconductors some unusual and useful properties.

As stated earlier, no electrons can lie above the Fermi level at absolute zero. However, higher temperatures add enough thermal energy to excite some electrons in a semiconductor above the Fermi level, from the valence band to the conduction band. Thus, although they are poor conductors, semiconductors can carry some current at room temperatures.

When an electron has enough energy to move into the conduction band, it adds a mobile electron to that band. This electron is free to migrate when a voltage is applied, but the electron's movement into the conduction band leaves an **electron hole** (usually just called a "hole") in the valence band. The hole acts like a particle with a positive charge. Thus, when a voltage is applied, a hole migrates toward the cathode, in the opposite direction of an electron.

This behavior brings up a question: Exactly what is a hole? It's certainly not a particle. It is, instead, the absence of a particle or,

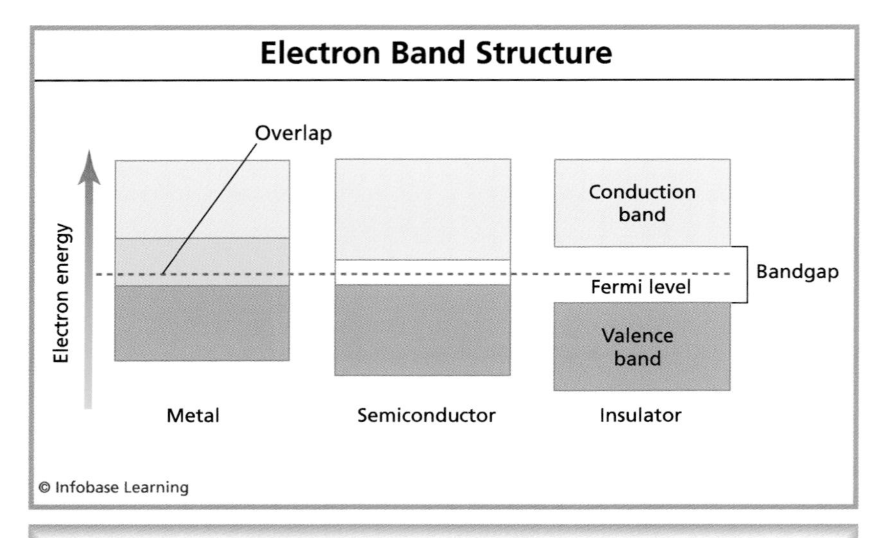

Figure 8.3 A simplified diagram of the electron band structure of metals, semiconductors, and insulators is shown.

more precisely, the absence of an electron. However, the missing electrons act just like particles.

Semiconductors come in two flavors: intrinsic and extrinsic. An intrinsic semiconductor is a crystalline solid, such as pure silicon or germanium. In an extrinsic semiconductor, a tiny amount (usually much less than 0.01%) of another substance is added to the silicon or germanium crystal. This process—known as doping—can have a dramatic effect on the conductivity of a semiconductor.

Both silicon and germanium have four valence electrons. When atoms with five valence electrons—such as antimony, arsenic, or phosphorous—are added to a crystal of silicon, they slip right into the crystalline lattice. However, each of them carries one more electron than their silicon neighbors. The additional electrons fill the crystal's higher-energy orbitals, which lie closer to the Fermi level. Thermal energy pushes some of the electrons into the conduction band, greatly increasing the conductivity of the crystal. Because doping with these elements adds negatively charged electrons to the semiconductor, they are called n-type semiconductors, where *n* stands for negative.

What would happen if you doped a semiconductor with holes? Of course, you cannot add a hole. However, you can add substances that create holes. Boron, aluminum, and gallium all have three valence electrons. When added to silicon (or germanium), these doping agents create a deficiency of electrons in the crystal lattice of the semiconductor. The missing electrons create holes and a type of semiconductor called a p-type semiconductor, where the *p* stands for positive, as in positive particle, which is how holes behave.

MA BELL AND THE TRANSISTOR

By the beginning of the twentieth century, Alexander Graham Bell's patents on the telephone were beginning to expire. This prompted Theodore Vail, the American Telephone and Telegraph (AT&T) Company's bold, new president, to consider how to keep the expected competition at bay. We will, he proclaimed, have "universal service." At that time, AT&T considered "universal" to mean the United States, and within five years, the company had made solid progress toward its goal of connecting the country.

To extend its reach, the company employed new technology. In 1906, the inventor Lee de Forest produced the first vacuum tube. It was a three-element device called an Audion. A wire grid was inserted between the negatively charged cathode and the positively charged anode. Tiny changes in power to the grid could produce big changes in the current flowing from cathode to anode. For example, a small negative voltage applied to the grid retarded the flow of electrons. Reversing the grid's polarity increased the flow. Thus, an Audion could amplify a signal.

The Audion and its successors proved useful for manufacturing radios, among other products. It could amplify a weak signal to provide the necessary power to operate headphones or speakers. It could also boost a telephone signal. Alexander Graham Bell illustrated that in 1915 when he repeated the famous command that he first issued almost 40 years earlier in the world's first phone call to a man in the office next to his lab: "Mr. Watson, come here. I want you." This time, though, Bell was in New York and Watson was in San Francisco.

De Forest's vacuum tubes did a good job of amplification, but they were bulky, balky, and hot. Scientists at "Ma Bell" (AT&T's nickname) were aware of the quantum revolution that was taking place in Europe. By the 1930s, that revolution was rewriting the laws of

Confusing Currents

The early scientists who experimented with the mysterious, invisible "stuff" called electricity knew that something was moving within their wires, something that could do work. Benjamin Franklin guessed that this moving force was positively charged. Thus, an electric current would flow from a positive pole to a negative one. Keep in mind that Franklin had no reason to call the force positive. It was simply a guess. Nobody knew what was moving.

After electrons were discovered in 1897, it became clear that Franklin had guessed wrong. The current in a wire is the movement of negatively charged electrons, which flow from negative pole to positive pole. The old system of positive to negative was so deeply entrenched, however, that engineers stuck with it. So, although we know today that electrons flow from negative to positive, most educational materials textbooks still depict an electric current as flowing from positive to negative, thereby confusing generations of students.

This is a less-than-ideal state of affairs. In this discussion, we will sidestep the problem by having electrons and currents flow in the same direction, from negative to positive.

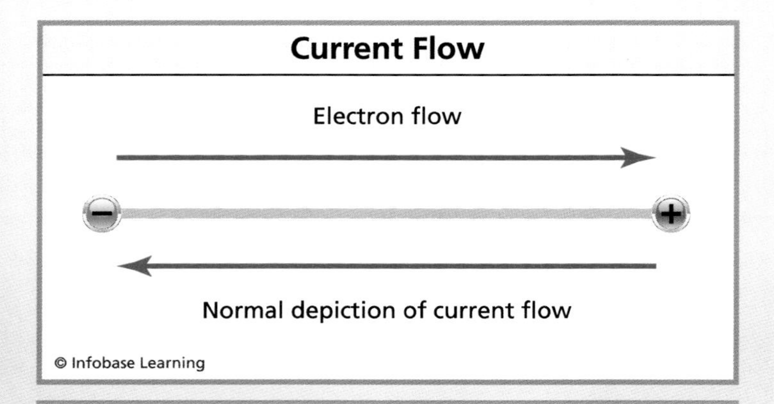

Figure 8.4 This is a conventional depiction of current flow.

solid-state physics. If vacuum tubes could be replaced by solid-state amplifiers—small, reliable, and cool—it would be a giant step forward for the company. Bell Labs—AT&T's research arm—decided to go into solid-state research in a big way.

The company's entry into the field was delayed by the onset of World War II, during which Bell Labs devoted much of its research to the war effort. Then, as the war wound down in 1945, physicist William Shockley and chemist Stanley Morgan were appointed to head a new solid-state physics group. Shockley was soon managing a small crew that included two outstanding scientists, Walter Brattain and John Bardeen. The group's job was to apply the quantum theory of semiconductors to produce a solid-state amplifier.

They had one good lead. Back in 1939, a Bell scientist named Russell Ohl had accidentally joined a p-type semiconductor to an n-type. The result was a semiconductor with a p-n junction, a circuit element to be examined later. This combination of semiconductors acted as a diode, a device that conducts electricity in one direction only. Ohl also found that under certain conditions, a p-n device could function as a solar cell, converting sunlight to electricity.

The next breakthrough came eight years later, just before Christmas in 1947. This time, it was not an accident that paved the way but impatience. Walter Brattain was working in the lab, following up on Ohl's discovery. He was studying the photovoltaic properties of a silicon crystal. Condensate, which would affect the outcome of his experiment, collected on the silicon, but instead of starting over again, Brattain submerged the entire experimental apparatus in organic solvents. "I'm a lazy physicist anyway," he declared. And a lucky one, too, because he was rewarded with a small photovoltaic effect. When he later dipped his apparatus in water, he saw a more pronounced photovoltaic effect. Furthermore, by varying the voltage across a pair of contacts inserted in the surface of the crystal, he could change the magnitude of the effect.

Working with Bardeen (but independent of Shockley), Brattain saw that the device had the potential to amplify a current. After a month, the two men produced the first solid-state amplifier. The device was named a "trans-resistor," which was soon shortened to transistor. Because of its design, the Brattain-Bardeen device was called a point-contact transistor. Although this discovery would win the two men a share of a Nobel Prize, the point-contact transistor

Figure 8.5 The discovery of the transistor was made by a trio of scientists at Bell Telephone Laboratories. Pictured are William Shockley (*seated*), who initiated and directed the Laboratories' transistor research program, as well as John Bardeen (*left*) and Walter Brattain (*right*).

was a skittish creature and never widely used commercially. A more important reason for industry's lack of enthusiasm for it was because, just after its discovery, William Shockley developed a more rugged and more reliable transistor.

The p-n Junction Diode

Here's why a semiconductor with a p-n junction functions as a diode. Connect the negative terminal of a battery to the n side of a p-n junction and a wire from the positive terminal

(continues)

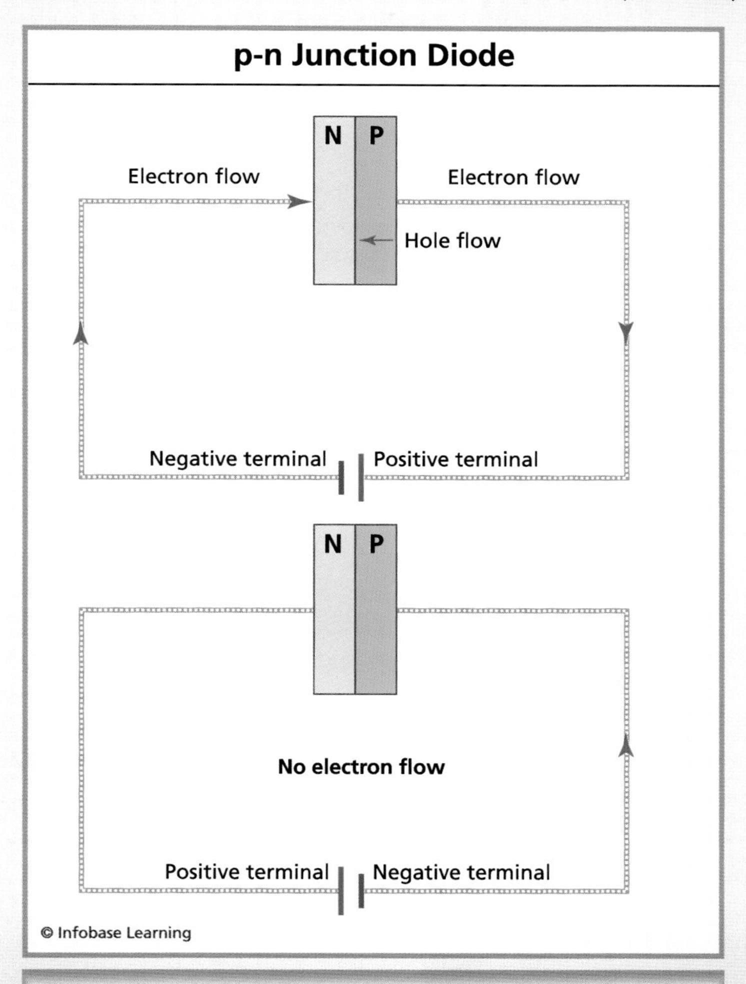

Figure 8.6 In electrical circuits with p-n junctions, negatively charged electrons move toward the anode, while positively charged holes would move toward the cathode. When the polarity is reversed, as in the bottom figure, the electrons cannot pass through the p-n junction.

(continued)

to the p side. The negative pole pushes electrons from the n side toward the junction. The positive pole pushes the holes in the p side toward the junction. This narrows the barrier at the p-n junction enabling the electrons to cross and move toward the positive terminal of the battery. The holes, acting as positive particles, move away from the positive terminal of the battery. Yet a hole moving in one direction is equivalent to an electron moving in the opposite direction. The net effect is that electrons on the p side move toward the positive terminal of the battery, and the diode conducts electricity.

When the flow of electrons is reversed, both holes and electrons are pulled away from the junction. The junction now acts as a barrier and little or no current can flow across it at ordinary voltages.

SHOCKLEY'S BIG IDEA

Shockley was agitated, sleeping poorly, and sick with disappointment following Brattain and Bardeen's triumph. The two scientists had beaten him to the first transistor. Shockley tried to get his name added to the patent application for the point-contact transistor. Although he had played no direct role in its development, he had contributed ideas, he said. No way, replied the two men, stunned at Shockley's gall. Rebuffed, Shockley decided to try to build a better mousetrap, a transistor superior to the point-contact device.

On January 23, 1948, Shockley succeeded. He came up with a new idea for a solid-state amplifier. The discovery would get him a share of the 1956 Nobel Prize in physics, along with Brattain and Bardeen. Like de Forest's vacuum tube, Shockley's transistor consisted of three components: A thin layer of p-type semiconductor sandwiched between two n-type semiconductors. Because of the arrangement of the layers, it is called an n-p-n transistor.

The elements of an n-p-n transistor are wired through a set of contacts. The contact to the thin layer of p-type silicon in the middle of the sandwich is called the base. The contacts with the n-type silicon are called the emitter and the collector.

With no current flowing to the base, little or no current will flow between emitter and collector. The reason is the p-n junction. Recall that a p-n junction will not pass electrons from the p side to the n side. However, if a small positive voltage is applied to the base, electrons will flow from emitter to collector.

Here's why: A positive voltage applied to the base pulls electrons from the emitter into the base. The electrons in the base migrate toward the positive pole of the power source at the collector. Thus, a small change in the voltage applied to the base will produce a large change in the current flowing from emitter to collector. Of course, a transistor (or any other amplifier) does not actually increase current. It simply allows more of the current from an external power supply to pass from emitter to collector.

Transistors can also act as switches. By changing or eliminating the voltage applied to the base, the current across the p-n junction can be started or stopped. This "current-on" or "current-off" feature can be used to represent and manipulate binary numbers, the language of digital computers. By treating current-off as zero and current-on as one, every transistor on a computer **chip** will represent a zero or a one. Knowing this, a skilled programmer can write word-processing software that will enable someone to type a book about the quantum theory—the science behind those switches.

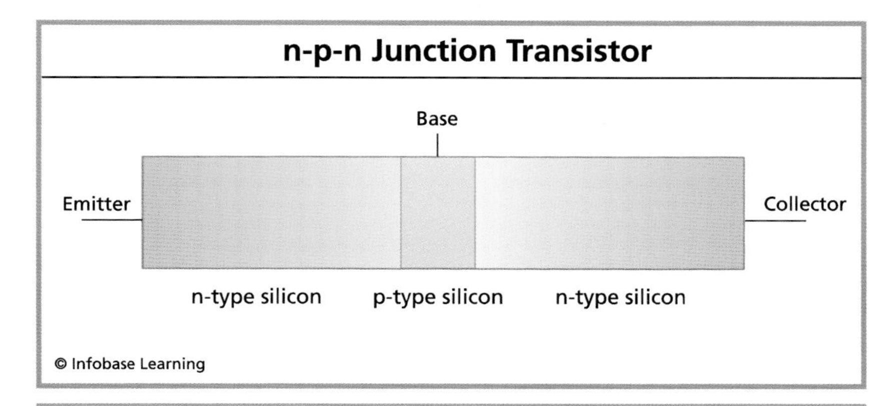

Figure 8.7 This is a schematic of an early transistor.

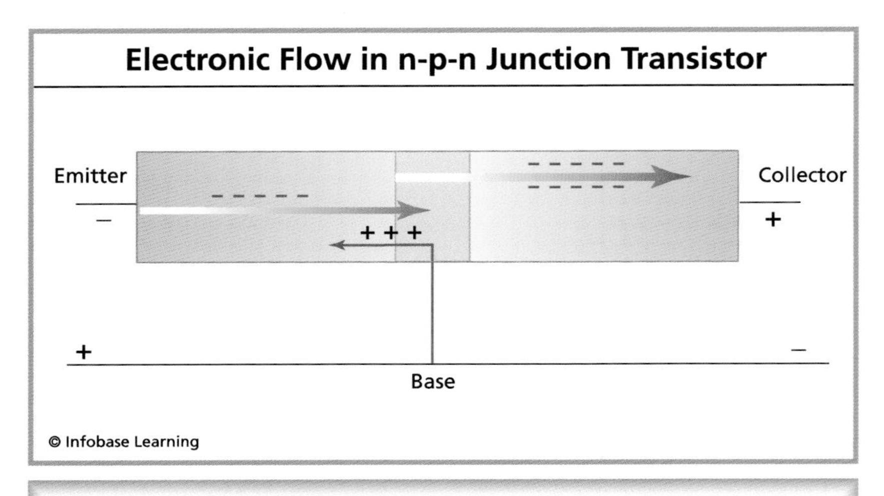

Figure 8.8 The base of the transistor acts as a gate that opens and closes in response to the voltage applied,

SO WHAT?

Now, back to the question posed earlier: Quantum theory—so what? Clearly, many features of modern life—from towering nuclear power plants to the microscopic transistors in computers—depend on our knowledge of the quantum. Understanding the basic principles of the theory helps one appreciate how the gizmos and gadgets of the modern world work. However, while this is a good reason to study the quantum, it is not the only one.

For centuries, the quantum lay hidden from science. Only in the twentieth century was this entirely unsuspected domain explored. Beneath the surface of our day-to-day world lies another weird and wonderful world. It does not operate like our human-sized world. It has its own rules and its own beauty, and the deeper you dig into this new world, the stranger it becomes. Bohr's comment is as true today as it was in the 1930s: "If quantum mechanics hasn't profoundly shocked you, you haven't understood it yet."

Fortunately, trying to develop that understanding is an exciting and enlightening journey that will lead those who persevere to a deeper appreciation of the elegant and marvelously complex universe we inhabit

Glossary

absolute zero The lowest possible temperature, –273° C (459.7°F)

alpha particles Helium nuclei composed of two protons and two neutrons that are emitted in some types of radioactive decay

angular momentum A measure of the intensity of rotational motion

angular momentum quantum number Governs the angular momentum of the electrons in an atom and determines the shape of atomic orbitals

anode The positively charged electrode in an electrolytic system

Aufbau principle The lowest-energy orbitals fill first when electrons are added to build successive elements in the periodic table.

blackbody A hypothetical body that absorbs all radiation that reaches it

cathode The negatively charged electrode in an electrolytic system

chip A small piece of semiconducting material, usually silicon, on which a circuit is embedded; a typical chip contains millions of transistors.

classical physics The physics developed before the great advances of the twentieth century that led to quantum theory and relativity

commute A mathematical expression commutes if its value is independent of the order in which the operations are carried out; for multiplication, an expression obeys the commutative law if A x B = B x A.

conservation of energy The total energy of an isolated system remains unchanged regardless of transformations taking place within that system.

correspondence principle The results one obtains using the quantum theory should approach the results obtained with classical physics as the size of the system increases. In other words, a new theory must give the same results as an older one in the domains where the older systems work.

diffraction The bending or spreading of waves as they pass through a small aperture or around a corner

eigenvalues A special set of scalar quantities associated with the wave equation; in certain eigenfunctions, the resulting eigenvalues represent the energy of a system.

electromagnetic waves Waves of pure energy—from low frequency radio waves to high energy gamma rays, with light waves in between—that propagate through a vacuum at 3×10^8 m/s

electron A negatively charged particle found outside the nucleus of an atom; free electrons are called beta particles.

electron holes A deficiency of electrons in a lattice that act like positively charged particles

Fermi energy (or Fermi level) The energy level in a solid near or above which electrons are available to conduct a current

force The agency (symbol F) that produces acceleration in a body; Newton's second law states that $F = ma$, where m is the mass of the body and a is the acceleration force F imparts to it.

ground state The lowest stable energy state of a system; the term is usually applied to atoms and molecules.

Hamiltonian A mathematical operator named for the nineteenth-century mathematician William Hamilton; in quantum theory, it is the operator associated with a system's energy.

Hund's rule Atoms in a higher total spin state are more stable than those in a lower spin state; so, when electrons are added to successive elements to build the periodic table, they fill different orbitals before pairing up.

inertia The property of an object that makes it resist changes to its motion

insulator Material that is a poor conductor; an electrical insulator is a poor conductor of electricity.

ion An atom that carries an electric charge due to the addition or removal of electrons.

kinetic energy The energy of motion; the classical equation for kinetic energy of a body is $mv^2/2$, where m is the mass of the body and v is its velocity.

laser Stands for "light amplification by stimulated emission radiation"; a stimulated-emissions device that produces monochromatic, coherent, collimated electromagnetic radiation.

magnetic quantum number The third solution to Schrödinger's wave equation produces the magnetic quantum number; it specifies how the *s, p, d,* and *f* orbitals are oriented in space.

mass A measure of the quantity of matter; on Earth, weight indicates the mass of an object.

matrices A set of quantities in a rectangular array enclosed in brackets; matrix algebra is a set of rules that govern the operations one can perform on such arrays.

momentum Linear momentum is the product of a body's mass and velocity; it is usually represented by the letter p, so $p = mv$; see also *angular momentum*

monochromatic light Electromagnetic radiation of a single wavelength in the visible range.

motion The change in position of body with time within a frame of reference; absolute motion is meaningless; only relative motion can be determined.

noble gases Unreactive elements with a filled outer shell of electrons

orbital A subdivision of an energy subshell where there is a high probability of finding an electron; an orbital can contain a maximum of two electrons.

oscillator Any object (such as an atom) that vibrates in a periodic manner

partial differential equations Equations containing partial derivatives; those functions in which one variable changes and the others remain constant

Pauli exclusion principle No two electrons in an atom can possess an identical set of quantum numbers.

photoelectric effect Electromagnetic radiation knocks electrons out of a metal causing a current to flow; Einstein used this phenomenon to show that light was quantized and came in energy packets called photons.

photon A particle with energy but zero mass; it represents a quantum of electromagnetic radiation.

positron The antiparticle of the electron; it has the same mass as an electron but carries a positive charge of the same magnitude as the electron's negative charge.

principal quantum number Specifies the main energy shells of an atom; it corresponds roughly to the distance between the nucleus and the orbital; its symbol is n.

proton The positively charged subatomic particle found in the nucleus of atoms

quanta The plural of quantum; it is the minimum energy required to change certain properties, such as the energy of an electron in an atom.

quantum numbers The four quantum numbers—principal, angular momentum, magnetic, and spin—represent solutions to the wave equation and their values govern the electron configuration of atoms.

quantum theory (or quantum mechanics) The modern method for predicting and understanding the behavior of the world at the atomic level; it postulates that energy is not continuous but comes in irreducible packets called quanta.

radioactive elements Elements capable of emitting alpha, beta, or gamma radiation

SI units The International System of Units, abbreviated SI from the French *Le Système international d'unités*; this form of the metric system is almost universally used in science.

scientific notation A method for expressing numbers in the form of exponents of 10, such as $10^2 = 100$, $10^3 = 1{,}000$, and $6{,}020 = 6.02 \times 10^3$

semiconductor A substance such as a silicon crystal that conducts electricity better than an insulator but not as well as a conductor; its conducting energy band and its valence band are separate but close together.

spin quantum number In an atom, every electron has a spin quantum number; spin can have only one of two possible values, usually designated as + or -. Although originally thought of as an electron spinning on its own axis, there is actually no precise physical characteristic associated with this quantum number.

stimulated emission The quantum mechanical event on which lasers are based; a photon interacts with an atom already in an excited energy state; under certain conditions, the incident photon causes the atom to emit another photon of the same wavelength.

subshell One or more orbitals with the same principal quantum number and angular momentum quantum number; the p subshell, for instance, has three individual orbitals.

transistor A solid-state amplifier

uncertainty principle First proposed by Werner Heisenberg; certain quantities, such as the position and momentum of a body, cannot be precisely determined simultaneously; the more you know about an object's momentum, the less you can know about its position, and vice versa.

Bibliography

American Institute of Physics. "Triumph of the Copenhagen Interpretation." Available on line. URL: http://www.aip.org/history/heisenberg/p09.htm.

———. "The Uncertainty Principle." Available online. URL: http://www.aip.org/history/heisenberg/p08.htm.

Drake, Stillman. *Galileo at Work: His Scientific Biography*. New York: Dover Publications, 1978.

Cathcart, Brian. *The Fly in the Cathedral: How a Group of Cambridge Scientists Won the International Race to Split the Atom*. New York: Farrar, Straus and Giroux, 2004.

CERN Courier. "Paul Dirac: a Genius in the History of Physics." Available online. URL: http://cerncourier.com/cws/article/cern/28693

Cline, Barbara Lovett. *Men Who Made a New Physics.* Chicago: University of Chicago Press, 1987.

Duck, Ian and E.C.G. Sudarshan. *100 Years of Planck's Quantum*. Singapore: World Scientific, 2000.

Everitt, Francis. "James Clerk Maxwell: A Force for Physics." Physics World. Available online. URL: http://physicsworld.com/cws/article/print/26527.

Explain that Stuff! "Transistors." Available online. URL: http://www.explainthatstuff.com/howtransistorswork.html.

Ferris, Timothy. *Coming of Age in the Milky Way*. New York: Anchor Books, 1989.

Feynman, Richard. *QED: The Strange Theory of Light and Matter*. Princeton, N.J.: Princeton University Press, 1985.

———. *Six Easy Pieces: Essentials of Physics Explained by Its Most Brilliant Teacher.* Cambridge, Mass.: Helix Books, 1963.

Florida Community College at Jacksonville. "Solvay Conference 1927." Available online. URL: http://mooni.fccj.org/~ethall/trivia/solvay.htm.

Fowler, Michael. University of Virginia. "The Photoelectric Effect." Available online. URL: http://galileo.phys.virginia.edu/classes/252/photoelectric_effect.html

Galilei, Galileo. *Dialogue Concerning the Two Chief World Systems*. New York: The Modern Library, 2001.

Gleick, James. *Isaac Newton*. New York: Pantheon Books, 2003.

Gribbin, John. *In Search of Schrödinger's Cat: Quantum Physics and Reality*. New York: Bantam Books, 1984.

Hardy, G.H. *A Mathematician's Apology*. Cambridge, U.K.: The University Press, 1940.

Haselhurst, Geoff and Karene Howie. "Quantum Theory: Max Born." Available online. URL: http://www.spaceandmotion.com/quantum-theory-max-born-quotes.htm

Hecht, Jeff. *Beam: The Race To Make the Laser*. Oxford, U.K.: Oxford University Press, 2005.

Heilbron, J.L. *The Dilemmas of an Upright Man: Max Planck as Spokesman for German Science*. Berkeley: University of California Press, 1986.

International Solvay Institutes. Available online. URL: http://www.solvayinstitutes.be/

Isaacson, Walter. *Benjamin Franklin: An American Life*. New York: Simon & Schuster Paperbacks, 2003.

———. *Einstein: His Life and Universe*. New York: Simon and Schuster, 2007.

Koertge, Nora, ed. *New Dictionary of Scientific Biography*. Farmington Hills, Mich.: Gale Cengage, 2007.

Kragh, Helge. Physics World. "Max Planck: the Reluctant Revolutionary." Available online. URL: http://physicsworld.com/cws/article/print/373

Lightman, Alan. *The Discoveries: Great Breakthroughs in 20th Century Science, Including the Original Papers*. New York: Pantheon Books, 2005.

Lindley, David. "Invention of the Maser and Laser." Available online. URL: http://focus.aps.org/story/v15/st4.

Mahon, Basil. *The Man Who Changed Everything: The Life of James Clerk Maxwell*. Chichester, England: John Wiley & Sons, Ltd., 2003.

Manning, Phillip. *Atoms, Molecules, and Compounds*. New York: Chelsea House, 2007.

Michon, Gérard P. "Final Answers: Quantum Mechanics." Available online. URL: http://home.att.net/~numericana/answer/quantum.htm

MIT Spectroscopy Lab. "History." Available online. URL: http://web.mit.edu/spectroscopy/history/history-quantum.html

Moore, Walter. *A Life of Erwin Schrödinger*. Cambridge, U.K.: Cambridge University Press., 1994.

Morton, Alan Q. *Science in the 18th Century: The King George III Collection*. London: Science Museum, 1993.

Nave, Carl R. "Fermi Level." Available online. URL: http://hyperphysics.phy-astr.gsu.edu/hbase/solids/fermi.html

———. "Laser Applications." Available online. URL: http://hyperphysics.phy-astr.gsu.edu/Hbase/optmod/lasapp.html

———. "Quantum Physics." Available online. URL: http://hyperphysics.phy-astr.gsu.edu/Hbase/quacon.html#quacon

Niagara College. "How Lasers Work." Available online. URL: http://www.technology.niagarac.on.ca/courses/tech238g/Lasers.html

Nobelprize.org. "Henri Becquerel." Available online. URL: http://nobelprize.org/cgi-bin/print?from=%2Fnobel_prizes%2Fphysics%2Flaureates%2F1903%2Fbecquerel-bio.html

———. "Marie Curie." Available online. URL:http://nobelprize.org/nobelprizes/physics/laureates/1903/marie-curie-bio.html

Pais, Abraham, Maurice Jacob, David I. Olive, and Michael F. Atiyah. *Paul Dirac: The Man and His Work*. Cambridge, U.K.: Cambridge University Press, 1998.

Parker, Barry. *Quantum Legacy: The Discovery That Changed Our Universe*. Amherst, N.Y.: Prometheus Books, 2002.

The Physical World. "The Clockwork Universe." Available online. URL: http://www.physicalworld.org/restless_universe/html/ru_2_11.html

Physicsworld.com. "Paul Dirac: the Purest Soul in Physics." Available online. URL: http://physicsworld.com/cws/article/print/1705

Preston, Diana. *Before the Fallout: From Marie Curie to Hiroshima*. New York: Walker & Co. 2005.

Reeves, Richard. *A Force of Nature: The Frontier Genius of Ernest Rutherford*. New York: Atlas Books, 2008.

Resonance: The Journal of Science Education. "Discovery of Radium." Available online. URL: http://www.ias.ac.in/womeninscience/Curie.pdf

Rhodes, Richard.. *The Making of the Atomic Bomb*. New York: Simon & Schuster, 1986.

Riordan, Michael and Lillian Hoddeson. *Crystal Fire: The Birth of the Information Age*. New York: W.W. Norton, 1997.

Rosenblum, Bruce and Fred Kuttner. *Quantum Enigma: Physics Encounters Consciousness*. Oxford, U.K.: Oxford University Press, 2006

Rubin, Julian. "Heinrich Hertz: The Discovery of Radio Waves." Available online. URL: http://www.juliantrubin.com/bigten/hertzexperiment.html

S.O.S. Mathematics. "Multiplication of Matrices." Available online. URL: http://www.sosmath.com/matrix/matrix1/matrix1.html

Spencer, Ross. "A Ridiculously Brief History of Electricity and Magnetism." Available online. URL: http://maxwell.byu.edu/~spencerr/phys442/node4.html

Spradley, Joseph. "Hertz and the Discovery of Radio Waves and the Photoelectric Effect." *The Physics Teacher* (November 1988) 492–497.

Susskind, Charles. *Heinrich Hertz: A Short Life*. San Francisco: San Francisco Press, 1995.

University of Chicago. "Introduction to Astronomy: The Strange World of Quantum Mechanics." Available online. URL: http://flash.uchicago.edu/~rfisher/saic/saic_spring08_02.pdf

University of Winnipeg. "The Photoelectric Effect." Available online. URL: http://theory.uwinnipeg.ca/physics/quant/node3.html

Watkins, Thayer. "Coherence in Stimulated Emissions." Available online. URL: http://www.sjsu.edu/faculty/watkins/stimem.htm

Westfall, Richard S. *The Life of Isaac Newton*. Cambridge, U.K.: Cambridge University Press, 1993.

Further Resources

Drake, Stillman. *Galileo: A Very Short Introduction*. Oxford, U.K.: Oxford University Press, 1980, reissued 1996.

Gamow, George. *Thirty Years That Shook Physics: The Story of Quantum Theory*. Garden City, N.Y.: Dover Publications, 1985.

Gamow, George. *Mr. Tompkins in Paperback*. Cambridge, U.K.: Cambridge University Press, 1969.

Townes, Charles H. *How the Laser Happened: Adventures of a Scientist*. Oxford, U.K.: Oxford University Press, 1999.

Web Sites

Egglescliffe School: Black Body Radiation
www.egglescliffe.org.uk/physics/astronomy/blackbody/bbody.html

This site offers an excellent introduction to blackbody radiation and the ultraviolet catastrophe. The language is clear and the illustrations useful.

National Taiwan Normal University, Virtual Physics Laboratory: Propagation of Electromagnetic Waves
www.phy.ntnu.edu.tw/ntnujava/index.php?topic=35

This is the place to go for a clear, animated illustration of how electromagnetic waves propagate.

1921 Spark Gap Transmitter/Ham Radio
www.youtube.com/watch?v=6ybYxkt3UxU

This site shows in detail the spark-gap transmitter on a ham radio.

Public Broadcasting System's "Transistorized"
www.pbs.org/transistor/

PBS offers an easy-to-navigate site that shows the history of the development of the transistor and the integrated circuit.

Purplemath: Scalar and Matrix Multiplication
www.purplemath.com/modules/mtrxmult.htm

Not up on the details of matrix multiplication? This site—with matrices that fill themselves in—will remedy the problem.

Solvay Conference 1927
www.youtube.com/watch?v=8GZdZUouzBY

The titans of quantum theory came together at the Solvay Conference in 1927. These video clips show the greatest scientists of the day going about their business at the conference.

Picture Credits

Page

10: © Infobase Learning
15: © sciencephotos/Alamy
16: © Infobase Learning
17: © Infobase Learning
18: © SSPL via Getty Images
19: © Infobase Learning
21: © SSPL via Getty Images
29: © Infobase Learning
31: © Infobase Learning
33: © The Philadelphia Museum of Art/Art Resource, NY
35: © Infobase Learning
36: © Infobase Learning
37: © Pictorial Press Ltd/Alamy
39: © World History Archive/Alamy
42: © Infobase Learning
44: © Infobase Learning
45: © Infobase Learning
48: © Gamma-Keystone via Getty Images
50: © Infobase Learning
52: © Infobase Learning
61: © Segre Collection/AIP/Photo Researchers, Inc
62: © Infobase Learning
64: © Mary Evans Picture Library/Alamy
72: © Infobase Learning
73: © Infobase Learning
75: © Infobase Learning
81: AIP Emilio Segre Visual Archives
90: © Photos 12/Alamy
96: © Infobase Learning
99: © Infobase Learning
102: © Infobase Learning
104: © Infobase Learning
106: © INTERFOTO/Alamy
107: © Infobase Learning
109: © Infobase Learning
110: © Infobase Learning

Index

About the Author

Phillip Manning is the author of 6 other books and 150 or so magazine and newspaper articles. His most recent books are *Atoms, Molecules, and Compounds* and *Chemical Bonds*. An earlier book, *Islands of Hope*, won the 1999 National Outdoor Book award for nature and the environment. Manning has a Ph.D. in physical chemistry from the University of North Carolina at Chapel Hill. His Web site at www.scibooks.org lists new books and book reviews about science.

Manning was assisted in this project by **Dr. Richard C. Jarnagin** who taught chemistry at the University of North Carolina for many years. He mentored numerous graduate students, including the author. Manning notes: "In assisting with this book, he caught several errors. Regrettably, any that remain are entirely mine."